Plug-in Electric Vehicle Grid Integration

For a complete listing of titles in the
Artech House Power Engineering Series,
turn to the back of this book.

Plug-in Electric Vehicle Grid Integration

İslam Şafak Bayram

Ali Tajer

ARTECH HOUSE

BOSTON | LONDON

artechhouse.com

Library of Congress Cataloging-in-Publication Data
A catalog record for this book is available from the U.S. Library of Congress.

British Library Cataloguing in Publication Data
A catalogue record for this book is available from the British Library.

Cover design by John Gomes

ISBN 13: 978-1-63081-051-1

685 Canton Street
Norwood, MA 02062

10 9 8 7 6 5 4 3 2 1

Dedicated to my lovely wife Sinemis - Şafak

Dedicated to my family - Ali

Contents

Preface xiii

Chapter 1 Introduction to Plug-in Electric Vehicles 1
1.1 Introduction 1
1.1.1 Environmental Aspects 2
1.1.2 Cost Savings 3
1.1.3 Energy Independence and Security 3
1.2 History of Electric Transportation 5
1.2.1 Early Days 5
1.2.2 Decline of Electric Vehicles 5
1.3 Electric Vehicle Models and Specifications 8
1.3.1 Battery Electric Vehicles 9
1.3.2 Plug-in Hybrid Electric Vehicles 9
1.3.3 Hybrid Electric Vehicles 9
1.3.4 Fuel-Cell Electric Vehicles 10
1.3.5 Fuel Efficiency Measures 10
1.3.6 Comparison of Carbon Emissions 14
1.4 Plug-in Electric Vehicle Sales 17
1.5 Chapter Summary 20

Chapter 2 Enabling Technologies 23
2.1 Battery Technology 23
2.1.1 Lithium-ion Batteries 24
2.1.2 Basic Battery Terminology 25
2.1.3 Design Criteria 30
2.1.4 Battery Management Systems 31
2.2 Plug-in Electric Vehicle Drivetrain Architectures 32

2.3 Electric Vehicle Service Equipment 34
2.3.1 Battery Charger Technology 35
2.3.2 Charging Rates and Options 37
2.3.3 Wireless Charging 41
2.3.4 Charging Station Deployment 41
2.4 Chapter Summary 42

Chapter 3 Roadblocks to PEV Integration 45
3.1 Impacts of PEV Charging 46
3.1.1 Distribution Network 46
3.1.2 Impacts on Generation Portfolio 54
3.1.3 Transmission Network 55
3.2 Network of Charging Stations 55
3.3 High Acquisition Cost 57
3.4 Chapter Summary 59

Chapter 4 PEV Cost Models 61
4.1 Battery Degradation Cost 61
4.1.1 PEV Battery Modeling 62
4.1.2 Modeling Approach I: NREL Model 62
4.1.3 Modeling Approach II 67
4.1.4 Modeling Approach III 71
4.2 Electricity Cost 72
4.2.1 All-in-Rate 72
4.2.2 Time-of-Use Rates 74
4.2.3 Critical-Peak Pricing 74
4.2.4 Real-Time Pricing 75
4.2.5 Inclining Block Rates 75
4.3 Case Study 76
4.4 Chapter Summary 79

Chapter 5 Economics of Charging Services 83
5.1 Introduction 83
5.2 Charging for Services 85
5.2.1 Consumer's Perspective 86
5.2.2 Service Provider's Perspective 87
5.2.3 Socially Optimal Pricing 88
5.3 Other Pricing Constraints 89
5.3.1 Finite Capacity 89

5.3.2 Auction-based Pricing 90
5.3.3 Untruthful Behavior 90
5.4 Teletraffic Theory 91
5.5 Congestion and Its Effect on Pricing 91
5.6 Network Externalities 93
5.7 Demand Forecasting 94
5.8 Chapter Summary 95

Chapter 6 Capacity Planning for Large-Scale PEV Charging Stations 97
6.1 Introduction 97
6.2 System Description 99
6.3 Storage Capacity Analysis for Single-class PEVs 102
6.3.1 Energy Storage Access Dynamics 102
6.3.2 Analyzing Distributions 103
6.3.3 Single-charger Technology with One PEV 106
6.3.4 Single-charger Technology with Multiple PEVs 108
6.4 Storage Capacity Analysis for Multiclass PEVs 109
6.4.1 Effective Demand 110
6.5 Numerical Examples 111
6.5.1 Single-class Customers 111
6.5.2 Multi-class Customers 114
6.6 Chapter Summary 115

Chapter 7 Capacity Planning for Small-Scale PEV Charging Stations 119
7.1 Introduction 119
7.2 System Description 121
7.2.1 Stochastic Model 122
7.2.2 Toy Example 125
7.3 Profit Model 126
7.4 Performance Evaluation 127
7.5 Stations with Competing Energy Storage Technologies 133
7.6 System Metamodeling Approach 135
7.7 Chapter Summary 136

Chapter 8 PEV Demand Control 139
8.1 Introduction 139
8.2 Load Management for Fast-Charging Stations 141
8.2.1 Load Management at a Fast-Charging Station 141
8.2.2 Numerical Evaluation 145

8.3 Load Management at Multiple Charging Technology 146
8.3.1 System Model 147
8.3.2 Problem Formulation 148
8.3.3 Global Problem 149
8.3.4 Local Problems 150
8.3.5 Connection between Global and Local Problems 150
8.3.6 Computing LoLPs 151
8.3.7 Numerical Evaluation 156
8.4 Chapter Summary 158

Chapter 9 Optimal Service Rates 161
9.1 Introduction 161
9.1.1 Competing Performance Measures for PEV Charging 162
9.2 Optimal Individual Battery Charging 163
9.2.1 Numerical Evaluation 168
9.3 Charging Control at a Fast-Charging Station 169
9.3.1 Charging Station Queuing Analysis 173
9.3.2 Station Revenue Model 176
9.3.3 Numerical Evaluation 177
9.4 Chapter Summary 179

Chapter 10 Charging Facility Location Design 183
10.1 Introduction 183
10.1.1 Sample Objectives 186
10.2 Classification of Facility Siting Problems 188
10.2.1 Continuous and Discrete Network Models 188
10.2.2 Static and Dynamic Problems 191
10.2.3 Other Classification Methodologies 191
10.3 Siting Slow-charging Stations 193
10.4 Siting Fast-charging Stations 195
10.4.1 Case Study 200
10.5 Chapter Summary 202

Chapter 11 Load Balancing in Charging Station Networks 205
11.1 Introduction 205
11.2 Network of Fast-Charging Stations 206
11.2.1 PEV Admission Control 207
11.2.2 Resource Allocation Framework 210
11.3 Decentralized Control of Charging Stations 214

11.3.1 Game Formulation 214
11.3.2 PEV Drivers (Followers) 215
11.3.3 Network Operator (Leader) 217
11.4 Numerical Results 218
11.4.1 Spatial Vehicle Distribution 218
11.4.2 Simulation Setting 219
11.5 Chapter Summary 221

Chapter 12 Vehicle-to-X Systems 225
12.1 Introduction 225
12.2 Vehicle-to-Grid (V2G) 226
12.3 Power System Ancillary Services: An Overview 230
12.3.1 Regulation 230
12.3.2 Reserves 233
12.3.3 Motor Starting 233
12.3.4 V2G Demonstrations at PJM and CAISO Regions 234
12.4 Behind-the-Meter Applications 237
12.4.1 Vehicle-to-Building 237
12.4.2 Vehicle-to-Home 240
12.4.3 Vehicle-to-Vehicle 241
12.5 Chapter Summary 242

Chapter 13 Integration Beyond Technology: Social and Economical Aspects 245
13.1 Consumer Views on PEVs: Empirical Studies 246
13.2 Theoretical Frameworks for PEV Adoption 251
13.2.1 Diffusion of Innovations 251
13.2.2 Theory of Planned Behavior and Rational Choice Theory 254
13.3 Case Studies for PEV Promotion 255
13.3.1 Zones with Controlled Emissions 255
13.3.2 Biotaxis in Bogota, Colombia 256
13.3.3 E-Trike in Manila, Philippines 257
13.3.4 Greening Public Transport in Rome, Italy 257
13.3.5 Car-sharing in Paris, France 258
13.3.6 Workplace Charging Challenge, United States 258
13.3.7 Zero-Emission Vehicle Program, United States 258
13.4 Chapter Summary 259

List of Acronyms 263

About the Authors 265

Index 267

Preface

The United States National Academy of Engineering has identified electric systems and ground transportation as the two key engineering achievements of the twentieth century. While both domains constitute crucial infrastructures and have been successfully servicing a wide range of social and economical needs, they have evolved and developed in isolation, and the interactions between them have been limited. During the past decade, however, we have been witnessing an increasing interest in promoting mainstream adoption of the plug-in electric vehicles (PEVs) due to their various economical and environmental benefits. Specifically, the major drivers for this impetus include environmental concerns due to rising carbon emissions, the increasing dependency on crude oil and the undesirable impacts of fluctuating oil market conditions, and the shift towards modernizing the electricity grids and integrating renewable energy sources.

Integrating PEVs into power grids necessitates a holistic analysis due to the various complex interactions that it brings about at the interface of the two already complex networks (i.e., electricity and transportation systems). Forming various inferential, scheduling, and management decisions involves decision mechanisms that require an interdisciplinary approach. From the electric vehicle perspective, ensuring a desirable cruising range is of utmost priority. Facilitating long ranges necessitates electric motors for vehicle propulsion to replace internal combustion engines. The two main components of electric motors are onboard batteries for storing the required energy, and advanced power electronics that enable and control effective energy conversion. The interactions between the electric vehicles and the power grid are exerted by electric chargers, which should be designed to deliver the charging needs for various durations and PEV technologies. Chargers, when equipped with bidirectional energy transfer capabilities through their batteries, can also allow vehicles to inject power back to the power grid under certain circumstances. Due to such twofold roles, hence, vehicle batteries are the key enabling components of PEV technology. The capacity, various efficiency measures, and the cost of batteries are the primary factors that influence PEV sales and penetration rate.

All the aforementioned challenges reflect on the complex nature of a single PEV. When facing transportation systems with a vast numbers of PEVs, these challenges are further coupled with several others pertinent to resource management. Specifically, since vehicles will be fueled by the power grid, designing, planning, and operating the charging facilities that manage power transfer from the grid to vehicles will have instrumental roles. If not designed properly, such facilities can underutilize the grid assets or impose strenuous conditions on the grid. Addressing such challenges suggests that charging stations may need to be equipped with ancillary support units (e.g., energy storage units) to protect the grid from excessive customer demand, which in turn adds to the design complexity. Designing networks of charging stations is also coupled with other practical constraints such non-homogeneous customer density in different areas, in which case network operators may need to influence customers' choices by monetary incentives to defer or change their preferred locations of the service.

In addition to addressing the *technological* needs and the *economic* programs, the *human factor* should be also accounted for carefully. For instance, purchasing a PEV, for many individuals, requires changing driving routines, which might contribute to delaying or slowing down the mainstream acceptance even when PEVs are economically viable. Incentive programs focusing on the technology enthusiast may be successful at the early stages; however, mainstream acceptance requires social studies to be involved along with technological and economics studies. To that end, our approach in this book is three-dimensional and portrays a holistic picture about the challenges associated with mainstream acceptance of PEVs at three distinct layers: (1) technology layer, (2) economic layer, and (3) social layer. Chapters 1, 2, 3, 4, and 9 address challenges pertinent to the technology layer, Chapters 5, 6, 7, 8, 10, 11, and 12 focus on the issues in the economic layer, and Chapter 13 addresses challenges in the social layer. It is noteworthy that some of the chapters lie in the intersection of the three layers. For instance, Chapter 9 shows a profit optimization by considering a battery degradation of a vehicle. The details of each chapter are outlined next.

Chapter 1 provides an introduction to PEVs, discusses the advantages of PEVs over combustion engine vehicles, provides a historical background on PEVs, examines different types of PEVs, and provides an overview of current PEV models as well as market penetration tends. Chapter 2 is devoted to the enabling technologies, where the first half focuses on battery technology, and the second half is dedicated to drivetrain technologies and electric vehicle supply equipment. Chapter 3 concentrates on major roadblocks for PEV adoption and the negative impacts of PEV charging on the power grid. Chapter 4 focuses on PEV cost models (e.g., battery degradation models and different pricing mechanisms).

Chapter 5 is a transitional one that links the technology layer to the economic layer, and provides details of capacity planning, demand control, and charging rate control studies. Chapter 6 focuses on a capacity planning framework for single and multi-class customers in large-scale charging facilities, while Chapter 7 presents a study on capacity and energy storage sizing in the fast-charging station. Chapter 8 discusses pricing-based demand control in charging networks, and Chapter 9 is focused on controlling the charging rates in a fast-charging station. The topics and discussions presented up to Chapter 10 are entirely focused on settings that consist of only one charging station, while Chapter 10 considers optimal siting of a network of charging stations. Chapter 11 concentrates on a load-balancing framework in a charging network through pricing.

PEV batteries can also be used to support grid activities such as regulation, storage, and emergency backups. Motivated by this application, Chapter 12 is dedicated to vehicle-to-grid, vehicle-to-building, vehicle-to-vehicle, and vehicle-to-home energy transfer applications. The last part of the book, Chapter 13, is reserved for reviewing the social aspects involved in designing and operating PEV networks.

We would like to express our sincere gratitude to all those who helped and supported us to complete this book. Even thought they are too numerous to be listed, the authors are grateful to all. We thank the staff of Artech House Inc., and Aileen Storry for initiating this book project and her support during the time we work on the book. Last but not least, we thank the anonymous reviewers whose comments made this book better.

Chapter 1

Introduction to Plug-in Electric Vehicles

1.1 Introduction

The transportation industry has had a revolutionary role in various economic and social developments in the past century. These revolutions, often facilitated by enhanced mobility, have in turn fostered transformations in other sectors ranging from heavy industries to the access of individual citizens to medical care and educational systems. Such ubiquitous dependence on vehicles, therefore, has placed the invention and mass production of personal vehicles as a major technological achievements of the twentieth century [1–3]. The current rate of car ownership, which varies between 600 to 800 cars per 1000 citizens in the United States, Europe, and Australia, indicates that personal vehicles have become irreplaceable elements of our lives. Such pervasive presence of vehicles, as a result, necessitates transportation systems that operate efficiently and effectively.

Since the industrialized mass-production of vehicles, crude oil has been the primary source of fuel for urban transportation, as it is produced on a global scale, it is widely accessible, and its high energy density has maintained the cost of transportation at an economically affordable rate. Nevertheless, over time the advantages of crude oil have been plummeting due to the substantial increase in oil prices as well as increased awareness of the adverse environmental consequences of consuming fossil-based fuels. Motivated by addressing these issues, there has been an impetus towards evolving the transportation systems to a more economically and environmentally sustainable state, and electrifying the transportation systems has emerged as an attractive and viable solution. Electric transportation systems offer solutions to an array of issues, ranging from environmental concerns to the cost-effectiveness of fossil-based fuels driven by unstable oil markets.

Since their first introduction in late 2010, more than 800,000 *plug-in* electric vehicles have been deployed worldwide, and this trend is expected to grow well into the next decades. The multifaceted benefits of plug-in electric vehicles (PEVs), which will be presented next, have shaped a broad consensus about the necessity of transforming light-duty ground transportation from one that is gas-powered to one that is predominantly fueled by electricity. It is noteworthy that in this book we consider plug-capable vehicles, meaning that the vehicle is equipped with onboard batteries that can be charged by external energy sources such as power grids. A more detailed analysis of different vehicle types is presented in Section 1.3.

1.1.1 Environmental Aspects

The transportation sector is responsible for nearly one-third of the entire amount of energy consumed worldwide, and it is one of the main contributors to greenhouse gas (GHG) emissions.[1] Annual variations of the GHG emission levels caused by the transportation sector are depicted in Figure 1.1(a), and indicate a steady increase in pollutant gas emissions. If not ameliorated properly, such emissions can lead to acute environmental consequences. Such GHG emissions are primarily due to the emissions of internal combustion engine vehicles (ICEVs), and to a lesser extent due to the emissions of power generation, transmission, and distribution systems. Hence, electrification of the transportation system replaces the ICEVs with electric powers, and significantly reduces such emissions. The study in [4] shows that even without changing the current power generation portfolio in Europe, electric transportation offers 10% to 24% decrease in global warming potential.[2] [3]

Furthermore, the emissions of the power generation systems can be circumvented by planning and designing the network of PEV charging systems. For instance, charging PEVs with renewable resources, such as solar and wind power,

1 According to the United States Environment Protection Agency (EPA), 13% of the global GHG emissions were caused by transportation systems in 2004, and this rate increased to 27% in 2009.

2 The United States EPA defines the global warming potential (GWP) as a relative measure to compare the amount of heat trapped by the certain gas in question to the amount of heat trapped by the same mass of carbon dioxide.

3 Assuming a lifetime of 150,000 km for one PEV.

incurs substantially lower emission levels compared to relying on the power generated by fossil-based fuels. Therefore, intelligent charging mechanisms are instrumental to harnessing GHG emissions. More detailed analysis of GHG emissions is provided in Section 1.3.6.

1.1.2 Cost Savings

As shown in Figure 1.1(b), the average cost of owning a gas-fueled vehicle has followed an increasing trend since 1990. Furthermore, the operating cost of gas-fueled vehicles substantially exceeds that of electric cars. The primary reason for such discrepancy in the operating costs is that electric motors enjoy high efficiency, typically about 90%, while the efficiency of combustion engines hovers around 25% – 30%. Moreover, the cost of gasoline has been unstable and fluctuating substantially, while that of electricity generation and transmission has been stable worldwide. Besides these potential advantages, electric vehicle (EV) systems can also flexibly capitalize on the power consumption fluctuations, and charge their batteries when the power grid is less congested and the available power capacity is underutilized. The economic dimensions of the electric transportation are discussed in detail in Section 1.3.5.

1.1.3 Energy Independence and Security

For most nations, energy independence and security are some of the primary motivations to reform the transportation sector. Countries such as the United States, most countries in Europe, China, and Japan are the major consumers of the global oil market, which is not always stable. For instance, the United States consumes one-fourth of the global demand, and 94% of its transportation sector relies on imported oil. Countries with dependence on importing oil might also be facing fiscal pressure and their trade deficit is increasing due to the volatility of oil markets, which are often intertwined with other geopolitical dynamics. Moreover, global natural resources are being depleted at an unprecedented pace, while the global demand for oil, especially in the developing countries, is increasing. Electrifying transportation systems, therefore, strengthens energy security and independence as electricity in each country is often generated through a diverse set of domestic sources.

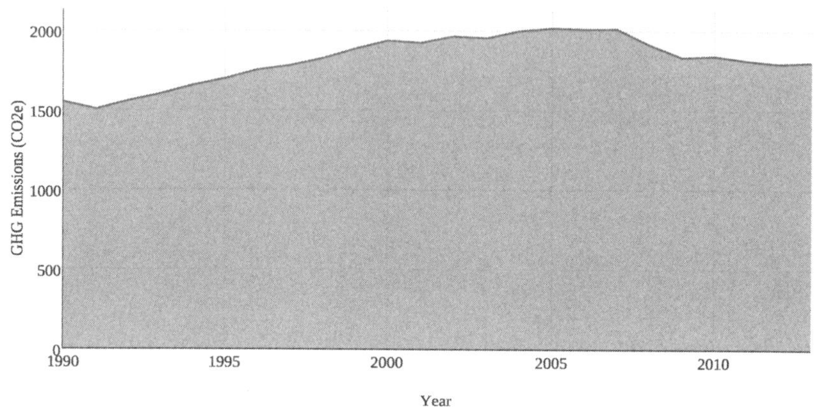

(a) GHG Emissions.

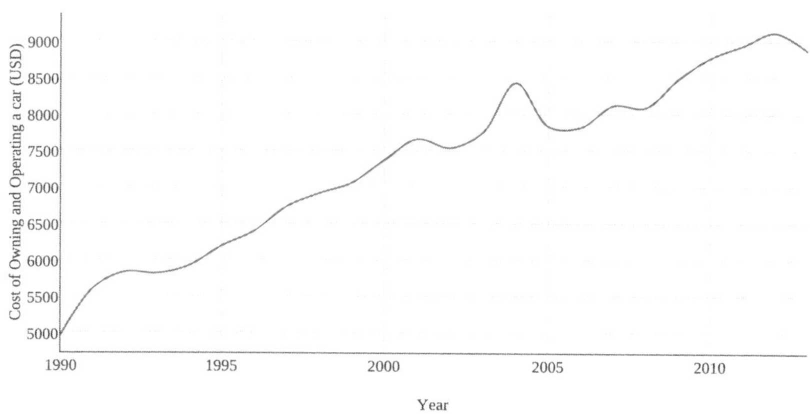

(b) Fuel cost.

Figure 1.1 The primary motivations for electric transportation are increasing emissions and fuel cost. (a) United States GHG emissions (carbon dioxide, methane, nitrous oxide, and fluorinated gases) from the transportation sector (Source: The United States EPA Inventory of United States GHG Emissions and Sinks. (b) Average cost of owning and operating an automobile in the United States.

1.2 History of Electric Transportation

1.2.1 Early Days

The history of the first electric-powered vehicles dates back to the 1830s when the Scottish inventor Robert Anderson designed the first vehicle with a non-rechargeable battery. In subsequent years, the interest in electric mobility grew and in 1859 Gaston Plante, a French physicist, invented the first commercial rechargeable battery, which enabled producing the first generation of EVs. The German engineer Andreas Flocken developed the first four-wheeled EV in 1888 and three years later, the first EV in the United States was introduced by William Morrison. In the 1890s, EVs attracted more interest, and a variety EV models were manufactured. In 1897, Pope Manufacturing Company in Connecticut, USA, became the leading large-scale EV producer, and its commercial products joined the taxi fleet of New York City. Another historical turning point in the past of EVs took place in 1899 when a Belgian car racer set a new land speed record of 106 kilometer per hour with his electric-powered vehicle known as *La Jamais Contente*.

At the beginning of the twentieth century, the market share of EVs grew to 28% of the total market in the United States. Some of the major manufacturers were the Electric Carriage and Wagon, Anthony Electric, Baker, Columbia Anderson, Edison, Riker, and Studebaker. The main advantage of EVs at that time was their more comfortable driving experience compared to their gas-power counterparts by having motors with reduced noise and vibrations. Also, driving an EV did not need changing gears, which was a significant shortcoming of gas-powered vehicles. Driven by these benefits, the global electric vehicle sales reached a historic peak of thirty thousand units in 1912.

1.2.2 Decline of Electric Vehicles

The popularity of EVs started to decline in 1920s due to a series of technological and economic developments that incentivized the deployment of gas-powered vehicles. Introduced by Henry Ford in 1908, the mass production of ICEVs made manufacturing gasoline-powered vehicles significantly more cost-effective compared to their previous generations. For instance, the price of a typical EV was 1750 USD, while the cost of a gasoline car was 650 USD. Furthermore, in 1912 Charles Kettering invented electric starters for ICEVs, which eliminated the need for hand cranks and made gas-powered vehicles more attractive. The discovery of crude oil in Texas further reduced the cost of transportation and gas-powered vehicles gained more

Table 1.1
Overview of Electric Vehicle History.

Year	Event
1832	Robert Anderson developed a non-rechargeable electric-powered two-wheel carriage in Scotland
1834	Thomas Davenport of the United States produced the first DC motor
1851	Wilhelm Sinsteden invented the rechargeable lead-acid battery
1859	Gaston Plante of France built the first commercially viable rechargeable lead-acid battery
1888	Andreas Flocking of Germany built the first four-wheeled electric vehicle
1898	Krieger Company of Electric Vehicles in France built EV models that had a top speed of 21 mph and 65 miles of all-electric range. This is the first EV with regenerative electric brakes
1899	Electric vehicle speed record broken
1912	Global peak for electric vehicle sales
1915	Woods Motor Company introduced the Dual Power that had both electric and internal combustion engine with a top speed of 35 mph
1930s	Interest in EVs dwindles
1960s	Environmental concerns sparked interest in EVs
1966	GM introduced Electrovair with a 115-hp, three-phase induction motor, silver-zinc battery, top speed of 80 mph, and an all-electric range of 40 to 80 miles
1973	OAPEC crisis raised concerns for energy independence
1993	GM introduced Impact 3 with an 137-hp, three-phase induction motor, 75 mph top speed, and 90 miles of all-electric range
2008	Oil prices reaches 150 USD per barrel
2010	Nissan introduces Leaf (leading environmentally friendly affordable family car) and in 2015 Leaf became the all-time best-selling all-electric vehicle

† From [5].

mainstream popularity. Besides these developments, the road system in the United States expanded significantly and the desire for interstate travel required facilitating longer driving ranges. Since public charging facilities were not widely available, battery technology became a bottleneck for widespread deployment of electric cars due to their low energy-to-weight return. For instance, energy density of an early lead-acid battery was 30 Wh/kg and even a 90% efficient motor yielded an energy density of 27 Wh/kg. On the other hand, the energy-to-weight ratio of an ICEV was 30 times higher than that of electric vehicles. As a result, even though the ICEVs suffered from low efficiency, typically around 25%, their useful energy density was 2250 Wh/kg. This simple example shows that in order to drive the same distance the required battery pack required was 84 times heavier than the required gasoline in an ICEV.

Due to the aforementioned issues, electric vehicles became all but extinct by the mid-1930s, and ICEVs prevailed over the competition. As a result, in the ensuing three decades the presence of EVs became significantly marginal. In the late 1960s more awareness was raised about environmental issues regarding ICEVs, which in turn again promoted the notion of electric transportation. Major manufacturers, such as Ford and General Motors, introduced two new vehicle models: Electrovair and Electrovan. In parallel, in the early 1970s, the instability in the oil market generated a new wave of interest in electric vehicles. Nations who went the oil embargo of 1973 considered electric transportation as an effective remedy to reduce dependency on importing oil. Even though this impetus was not strong enough to substantially increase deployment of electric vehicles, for the first time in its history, the United States Congress passed the Electric and Hybrid Vehicle Research, Development and Demonstration Act of 1976 (#94 – 413), which provided funding for research and development activities related to advancing EV-related technologies. Such efforts motivated the United States Post Office Department to gradually replace its fleet with electric cars. In 1973 the post office of Cupertino, California, became the pioneer in adopting the Harbilt electric vehicles for its entire fleet.

The interest in electric vehicles, however, was halted during 1980s and sparked again only after the State of California enacted the zero-emission vehicle mandate. In 1996, General Motors introduced EV1, which is considered the first mass-produced vehicle for modern times. However, the EV1 program was cancelled in 2002. At the same time, in 1997 Toyota introduced its Prius model as a hybrid car model equipped with a small battery pack, which is charged during acceleration. This model, while not being plug-capable, offered a major leap forward in the electric vehicle industry. An overview of the history of electric vehicles is presented in Table 1.1.

1.3 Electric Vehicle Models and Specifications

Electric vehicles are broadly categorized into four groups based on the electric design of their powertrains, namely battery electric vehicles (BEVs), plug-in hybrid electric vehicles (PHEVs), hybrid electric vehicles (HEVs), and fuel-cell electric vehicles (FCEVs). In this section the primary focus is placed on the plug-in vehicles, in which batteries are recharged by the electricity grid, and therefore, they affect other operations of the grid. BEVs and PHEVs fall in this category, which we collectively refer to as plug-in electric vehicles (PEVs). We also provide a brief overview of HEVs and FCEVs and compare all four models [6–10].

In order to promote a mainstream adoption of PEVs, some policy-making entities have implemented programs that offer incentives for deploying PEVs by providing expected specifications for the PEVs. For instance, the North Carolina State Senate describes a PEV as a four-wheeled motor vehicle that meets all of the following requirements:

1. It is made by a manufacturer primarily for use on public streets, roads, and highways.
2. It has not been modified from the original manufacturer specifications.
3. It is rated at no more than 8500 pounds unloaded gross vehicle weight.
4. It has a maximum speed capability of at least 65 miles per hour.
5. It draws electricity from a battery that has all of the following characteristics:
   (a) A capacity of not less than 4 kilowatt hours.
   (b) Capable of being recharged from an external source of electricity.
   (c) Acquired by the taxpayer on or after October 1, 2010.

In another example in the United Kingdom, the Office of Low Emission Vehicles has identified the following specifications for PEVs:

1. The vehicle must emit less than 75 grams of carbon dioxide per kilometer.
2. It must be able to travel a minimum of 70 miles between charges and have a minimum electric range of 10 miles.
3. It must be able to reach a speed of 60 miles per hour or more.

More details on the incentives will be presented in Chapter 3.

1.3.1 Battery Electric Vehicles

BEVs, also known as pure electric or full electric vehicles, are powered by an electric motor and do not employ an internal combustion engine. The battery of a BEV is charged primarily by the power grid and partially via regenerative breaking. Even though air pollution occurs during electricity generation, BEVs do not have tailpipe emissions and, therefore, they are also called zero-emission vehicles. In order to facilitate reasonable driving ranges, the BEVs typically have the largest battery packs compared to other models. The size of these battery packs usually varies between 20 kWh to 85 kWh with all-electric-range (AER) of $80 - 90$ miles. These ranges are selected by taking into account that 90% of the population in the United States travel less than 100 miles per day. The daily driving ranges are substantially lower in Europe and vary in the range $9 - 25$ miles in different countries. Hence, such battery packages can provide sufficient energy to drive BEVs for one or more days. Currently, a variety of commercial battery electric vehicles are available in the market, an overview of which is presented in Table 1.2.

1.3.2 Plug-in Hybrid Electric Vehicles

PHEVs employ both an electric motor and an internal combustion engine, which can use gasoline or diesel. Typically, PHEV batteries are smaller than the those of BEVs, because the required driving range is mostly supported by internal combustion engines and their combined driving range is higher than the that of BEVs. As a result, the main advantage of PHEVs is employing both engines types, which helps the consumers overcome range anxiety. For example, the Chevrolet Volt can be driven for 38 miles with the electric motor and the driving range can be extended for additional 350 miles with the internal combustion engine. In PHEVs, the electric motor is used mostly in urban environments where driving involves frequent stops. On the other hand, an internal combustion engine is used when the battery is depleted, during intensive cooling or heating, or during rapid acceleration. An overview of current PHEV models is shown in Table 1.3. The PHEV technology offers a clean and affordable transportation option for traveling long distances and at the same time ensuring reduced carbon emissions.

1.3.3 Hybrid Electric Vehicles

HEVs are primarily powered by internal combustion engines, which use conventional fuel sources such as gas and diesel. These vehicles are also equipped with battery packs, which are charged by regenerative breaking systems and are mainly used to enhance the fuel efficiency of the vehicle. The first generation of HEVs was presented to the market in 1997, when Toyota introduced the Prius model, and two years later Honda introduced the Insight model. Due to low gas prices, the full extent of the advantages of HEVs was not fully observed and these vehicles did not initially show the full extent of their potential. Nevertheless, energy prices surged in the following decade and HEVs emerged as more economically attractive options. As of mid-2015 more than 10 million HEVs were deployed worldwide. Due to design of their electric components, HEVs are not plug-capable, and hence their deployment does not have any major effect on power grids.

1.3.4 Fuel-Cell Electric Vehicles

FCEVs use fuel cells, typically hydrogen, to power their onboard electric motors or charge their storage units. The propulsion system in FCEVs is more efficient than those in the conventional ICEVs, and they do not emit any pollutants. FCEVs can be fueled within 10 minutes and they can drive up to 300 miles. Over the last few years, a number of car manufacturers and government agencies have supported research and development activities that accelerate fuel-cell technology. Unlike BEVs and PHEVs, the FCEVs do not use power grid to charge their batteries. Also, their onboard batteries are fairly small in size compared to those of BEVs, PHEVs, and HEVs. The early car models adopting this technology include Hyundai Tucson, Honda Clarity, and Toyota Mirai. The disadvantages of FCEVs included having limited networks of hydrogen fueling stations,[4] and the technology being still in its infancy stages.

1.3.5 Fuel Efficiency Measures

Fuel efficiency is one of the most important aspects of PEVs and accounts for energy consumption rate for a unit of traveled distance. For ICEVs, miles per gallon (mpg) and kilometers per liter are commonly used to represent fuel efficiency rates of vehicles. By incorporating the energy consumption of the PEVs for converting electric energy into motion, miles per gallon equivalent (MPGe) is often referred to as the effective efficiency of the PEVs. MPGe is a measure of distance traveled per unit energy and is calculated as follows. According to the United States Environmental Protection Agency (EPA), one gallon of gas has 115,000 BTUs of energy, which is the equivalent of 33.7 kWh of stored electricity. Therefore, 1 MPGe is

4 Only 10 hydrogen fueling stations were in existence in the United States as of 2014.

Figure 1.2 A sample Monroney label for a BEV.

approximately the equivalent of 1 mile per 33.7 kWh or 0.029674 miles per kWh. The US EPA calculates the battery-to-wheel MPGe based on the following formula:

$$\text{MPGe} = \frac{E_G}{E_M \times E_M} = \frac{33705}{E_M}, \tag{1.1}$$

where E_G is the energy content per one gallon of fuel, E_M is wall-to-wheel electrical energy consumed per mile (Wh/mi), which is determined by EPA, and E_E is energy (BTU) per kWh and is equal to 3.412. This methodology does not consider well-to-wheel life cycle (i.e., generation and transmission of electricity) in order to enable comparing PEVs with ICEVs on battery-to wheel and tank-to-wheel bases.

In Tables 1.2 and 1.3, the MPGe values of different BEVs and PHEVs are presented. The associated calculations assume a mixture of 45% of highway driving and 55% of urban driving. For instance, Nissan Leaf's fuel economy is 126 MPGe and 101 MPGe in urban and highway driving, respectively, and their weighted average is 114 MPGe. In the United States, publishing such information is mandatory for all new vehicles and is provided on the Monroney Label' prepared by EPA. An example of such labels is depicted in Figure 1.2, important elements of which are discussed next.

Vehicle Type (#1) This field shows the vehicle technology and fuel type, which include gasoline vehicle, diesel vehicle, compressed natural gas vehicle to represent EPA, hydrogen fuel cell vehicle for FCEVs, electric vehicle for BEVs, and plug-in hybrid vehicle electricity-gasoline for PHEVs.

Fuel Economy (#2) The fuel economy field shows the test results for city, highway, and combined MPG or MPGe values. As mentioned before, combined MPGe value is calculated by taking the weighted average of the city (55%) and the highway (45%) values. It is noteworthy that the city MPGe value for PEVs exceeds the highway MPGe, as more frequent regenerative braking enables more frequent opportunities for the batteries to be charged.

Fuel Economy Comparison (#3) The text in this field shows the category of the vehicle and compares the vehicle with all others in terms of fuel economy. For instance, Nissan Leaf is among the most fuel-efficient vehicles in the class of midsize vehicles, and Volkswagen e-Golf ranks among the best in compact cars. Comprehensive lists comparing the fuel efficiency of different makes and models are prepared by the EPA on an annual basis.

5 Years Fuel Cost (#4) This is one of the most important parts of the label, which shows the relative fuel savings in the first five years of the vehicles lifetime with respect to the average new vehicles. The five-year cost calculations assume that vehicle travels 15,000 miles per year and rely on five-year projections about fuel prices. For instance, in the United States an average utility price of 12 cents per kWh is used for these calculations.

Fuel Consumption Rate (#5) Fuel consumption rate shows the amount of energy used to travel 100 miles. For most PEVs, this is an estimated value since the actual AER can be less than one hundred miles.

Estimated Annual Cost (#6) Annual fuel cost estimation assumes that the vehicle travels 15, 000 miles per year and takes the projected electricity prices into account. For the BEV given in Figure 1.2, the vehicle is assumed to consume 34 kWh of energy in every 100 miles, based on which the annual energy consumption adds up to 5.1 MWh. Multiplying the total energy by the unit price leads to an estimated cost of $600 per year.

Fuel Economy and Greenhouse Gas Emissions (GHG) (#7) This field rates the vehicle from 1 to 10, representing worst to best fuel economy and GHG emission in ascending order. PEVs do not have tailpipe emissions, while having varying degrees of emissions due stemming from electricity

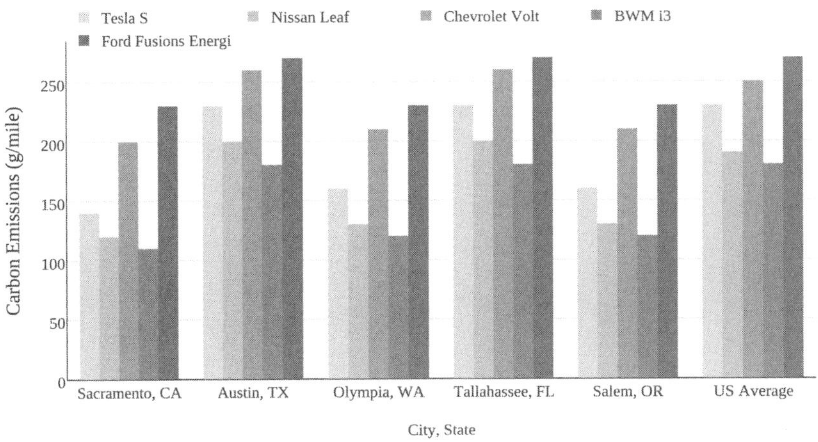

Figure 1.3 Carbon dioxide emissions per mile (g/mile) due to electricity production. Top 5 PEV adopter states and top 5 best-selling PEVs in 2015. Average new ICEV has an emission of 480 g/mile. Data from the United States EPA online emission calculator tool [11].

generation. Figure 1.3 compares the top five best-selling PEVs in 2015 in the capitals of the top five states in the United States with the most number of PEVs. It is noteworthy that Tesla S, Nissan Leaf, and BMW i3 are BEVs; therefore, they always perform better than the two presented PHEVs models (i.e., Chevrolet Volt and Ford Fusion). It is also noteworthy that in regions where pollutant generators such coal generators are deployed (e.g., Texas), carbon emission level corresponding to driving one mile is higher those of other regions. For instance, Sacramento, CA generates most of its power through clean generation sources such as hydroelectric plants (11%), renewable resources (22%), and nuclear power (22%), and as a result imposes lower carbon emissions. By comparing these results with the nation wide average emission level of 480 g/mile, it is observed that deploying BEVs leads to about a threefold reduction in emissions and the PHEVs can lead up to twofold such reductions. In the next section, we provide a more detailed analysis on the GHG emissions.

Table 1.2
Overview of Battery Electric Vehicles in 2015

Model	AER (mi)	MPGe Combined City/HWY	Engine (kW)	Battery (kWh)	5-year Cost Savings (USD)
BMW i3	81	124	125	21	6000
Chevrolet Spark	82	119	104	19	6000
Fiat 500e	87	116	83	24	5750
Ford Focus	76	104	107	23	5500
Kia Soul	93	105	50	16.4	5500
Mercedes B-Class	85	84	132	28	5000
Mitsubishi i-MiEV	62	112	49	16	5750
Nissan Leaf	84	114	80	24	5750
Smart Fortwo	68	107	55	17.6	5750
Tesla Model S	265	95	300	60	5250
Volkswagen e-Golf	83	116	85	24.2	5750

Smog Rating(#8) Smog rating shows vehicle tailpipe emissions leading to air pollution. In the United States standards, nitrogen oxide, carbon monoxide, non-methane organic gas, particulate matter, and formaldehyde are taken into account for determining this measure. For BEVs, the smog rating is 10, as there is no tailpipe emission.

Driving Range (#9) Driving range shows the approximate distance in miles that can be traveled with a fully charged battery.

1.3.6 Comparison of Carbon Emissions

The transportation industry is responsible for one-third of the total energy consumption and vehicles are its main contributors to carbon emissions. Electrification of

Table 1.3
Overview of Plug-in Hybrid Electric Vehicles in 2015

Model	AER (mi)	MPGe Combined City/HWY	Engine (kW)	Battery (kWh)	5-year Cost Savings (USD)
Cadillac ELR	37	82	111	17	3750
Chevrolet VOLT	38	98	118	17	4250
Ford C-Max	20	43	83	7.5	5500
Ford Fusion	20	88	68	7	4000
Honda Accord	13	115	124	7	4750
Porsche Cayenne S	14	47	70	11	500
Toyota Prius	11	95	60	4	4750
Porsche Panamera	16	50	70	9.4	750

transportation offers a unique potential for reducing carbon emissions. In this section, we provide a comparative analysis of vehicle emissions in different technologies by comparing BEVs, PHEVs, HEVs, and ICEVs. Generally, vehicles produce two types of emissions, namely air pollutants such as haze and smog, and greenhouse gasses such as methane and carbon dioxide. As mentioned earlier, PHEVs, when running in their gas mode, HEVs, and ICEVs deploy combustion engines and, consequently, they produce tailpipe emissions of both types. However, BEVs and PHEVs, when running in their battery modes, do not cause direct emissions, as they do not have tailpipes. To that end, a fair comparison requires well-to-wheel analysis that includes all emissions produced in all stages of production, transmission, distribution, and consumption of the fuel.

Emissions by the BEVs are governed by the power generation portfolio, since different fuel sources lead to different levels of emissions. For instance, production of one GWh electric energy by burning coal causes approximately 900 tons of carbon dioxide, while producing the same amount by hydroelectric plants causes only 26 tons of carbon dioxide. Furthermore, there can be differences in carbon emissions even with the same fuel types due to different levels of energy intensities, site-specific factors such as efficiency and the age of the power plants, as well as different processing techniques. Therefore, most of the power plants have their

Table 1.4
Statistics for Greenhouse Gas Emissions of Different Generation Options. †

Emissions (CO2e/GWh)	Coal	Oil	Natural Gas	Solar PV	Nuclear	Hydroelectric	Wind
Low	756	733	362	13	2	2	6
Mean	888	547	499	85	29	26	26
High	1310	935	891	731	237	237	24

† Source: World Nuclear Association [13].

own unique emission rates. In Table 1.4, we present statistics of the emissions in different generation types, which are based on various available studies. Argonne National Laboratory has devised *The Greenhouse Gases, Regulated Emissions, and Energy Use in Transportation Model* (GREET) [12], which is a software platform that determines gas emissions with high accuracy by using the up-to-date data about different generation options. Specifically, GREET uses the Powertrain System Analysis Toolkit (PSAT), a tool developed by Argonne National Laboratory, to simulate fuel economy and electricity usage of vehicles and to carry out wall-to-wheels analysis. We use GREET to analyze two representative settings in order to compare different models.

In the first setting, we choose and compare the emissions in the state of Oregon, in which half of the entire state's power is produced by hydropower, while coal constitutes less than one-third of the generation. As seen in Figure 1.4, the emissions caused by PEVs (both BEVs and PHEVs) are about one-third of ICEVs. Thus, PEVs can be regarded as an environmentally cleaner transpiration option. In the second setting, we analyze the state of Illinois, in which coal is the predominant power generation source. As shown in Figure 1.5, PHEVs and BEVs is smaller than its counterpart in Oregon due to the high percentage of coal usage. It is noteworthy that the calculations are carried out by taking the weighted average of each fuel type, hence showing average values. This also means that exact emission levels depend on the time of the day that the vehicles are charged due to the fact that energy dispatch has temporal variations during the day. For instance, at night time mostly nuclear generators are used, and as a result charging the vehicle during the night

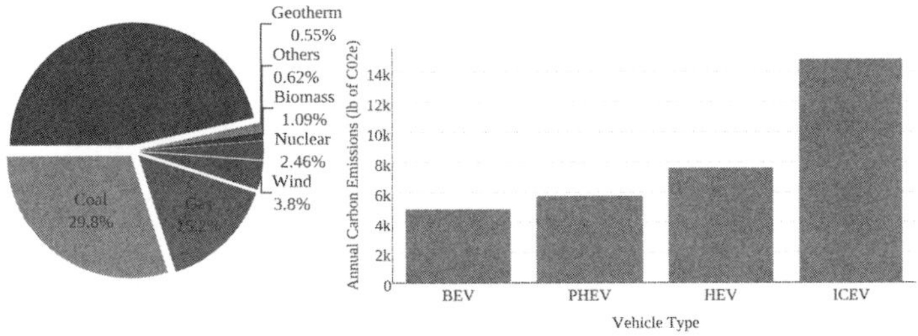

(a) Electricity generation mix. **(b)** Well-to-wheel emissions.

Figure 1.4 Comparison of well-to-wheel vehicle emissions in Oregon [11].

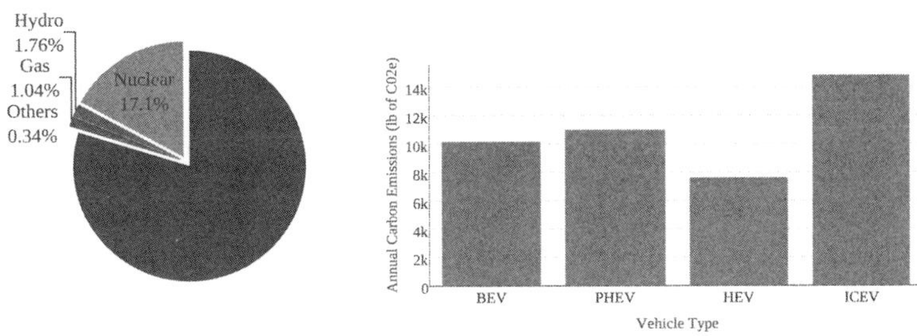

(a) Electricity generation mix. **(b)** Well-to-wheel emissions.

Figure 1.5 Comparison of well-to-wheel vehicle emissions in Springfield, Illinois.

causes less pollution than charging it during peak hours, during which gas-fired turbines are used. Comparing these two examples proves that in order to enjoy the environmental benefits of PEVs, clean generation options should be coupled with planning and devising proper PEV charging systems.

1.4 Plug-in Electric Vehicle Sales

Introduction and mass-production of various PEV models by major car manufacturers have expanded consumers choices drastically and have generated more public interest in adopting PEVs. Such trends in PEV penetration are expected to grow well

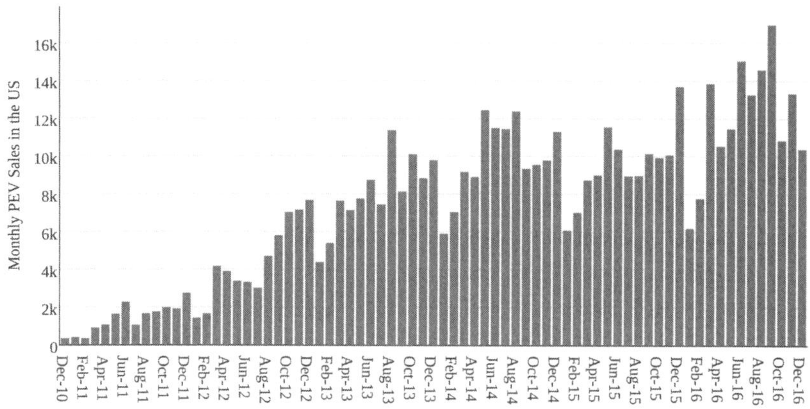

Figure 1.6 PEV sales in the United States [14].

into the future as various regulatory agencies and governments offer more incentives for adopting PEVs. Also, at the same time the public is showing a greater desire for more sustainable transportation systems. The penetration rates across the globe vary significantly due to a variety of factors including gas prices, average driving ranges, environmental concerns, government incentives, and the public interest. For instance, the price of one gallon of gasoline in Europe varies between 6 to 7 USD, while the same volume of gasoline costs about half that price in the United States. Moreover, transportation costs in Europe and Asia constitute a higher portion of an average household income compared to the United States.

As of December 2016, more than 2 million PEVs (including electric vans) have been sold worldwide. The PEV market is dominated by China, the United States, and Europe, and the market shares change slightly each year. For instance, in 2015, the United States lead the sales by having 41% of the market, followed by Japan with 15% and China with 12%. Figure 1.6 presents the monthly PEV sales in the United States. The sale rates in 2015 dramatically exceed the historical rate of 130,000 PEVs sold in 2014, and the same trend is expected to rise in subsequent years.

Besides gross sales statistics, sales per capita and percentage of new car sales also provide additional insights. For instance, Norway leads the global PEV

Table 1.5
Ranking of PEV Sales with Respect to the Share of New Car Sales

Country	Best-Selling Model	Total Sales (Q1 2014)	Total Sales (Q1 2015)	Share of Total Registrations
Norway	VW e-Golf	5,775	8,112	22.9%
Netherlands	VW e-Golf PHEV	3,306	5,760	5.2%
United Kingdom	Mitsubishi Outlander PHEV	1,764	8,684	1.2%
United States	Tesla Model S	14,799	14,832	0.8%
France	Renault ZOE EV	1,801	3,626	0.8%
Germany	VW e-Golf PHEV	2,286	4,520	0.6%
Japan	Nissan LEAF EV	9,626	7,750	0.6%
China	BYD Qin PHEV	1,486	12,555	0.3%

market, since more than one-fifth of new car sales are PEVs. Detailed data for the first quarter of 2015 is presented in Table 1.5. Based on these statistics, Europe is pioneering in adopting PEVs and five out of the eight listed countries with the highest penetration rate of PEVs are in Europe. Furthermore, most of the customers have a tendency towards their domestic or local manufacturers. For instance, Volkswagen is the best-selling vehicle in Germany, Norway, and the Netherlands, while Tesla S is the best-selling one in the United States market. Similar behavior is also observed in France, Japan, and China.

Even though there exist different market predictions for PEVs, there is a general consensus that PEVs will represent a sizable portion of new car sales in the next decade. For instance, Navigant Research estimates that global PEV sales will be around 5.8 to 7.0 million by 2023. One drawback of such predictions is that there exist conflicting factors that govern the penetration rate of PEVs. Specifically, customers' acceptance depends heavily on the cost savings and return on investment (ROI) periods. These two parameters are prone to volatile gas prices and international political climates, and can affect PEV markets adversely when the energy market is unstable. In the meantime, environmental regulations and motivations promote PEV adoption. For instance, as of 2015 in the United States, eight states– California, New York, Oregon, Maryland, Vermont, Rhode Island, Connecticut, and Massachusetts–, are aiming to have 3.3 active million PEVs replace gas-powered vehicles.

1.5 Chapter Summary

- Environmental concerns, volatile energy markets, and energy security are the main drivers for electric transportation.
- Electric vehicles were invented and developed in the nineteenth century but did not gain the mainstream acceptance because of their limited driving ranges, inadequate charging infrastructures, and the high cost of their batteries.
- Electric vehicles can be divided into four categories: battery electric vehicles (BEVs), plug-in electric hybrid vehicles (PHEVs), hybrid electric vehicles (HEVs), and fuel-cell electric vehicles (FCEVs). Only BEVs and PHEVs are plug-capable, and are referred to as plug-in electric vehicles (PEVs).
  - Technology maturity of HEVs dominates those of PHEV and BEVs, which in turn dominate that of FCEVs.
  - All electric range can be ordered as FCEV, BEV, PHEV, from the highest to lowest, and HEVs have zero all-electric range.
  - Battery sizes can be ordered as BEV, PHEV, HEV, and FCEV, in descending order.
  - Fuel savings can be ordered as BEV, PHEV, HEV, and FCEV, in descending order.
- In this book, we consider vehicles that are plug-capable (i.e., BEVs and PHEVs) and collectively refer to them as PEV (PEVs). The advantages of PEVs over ICEVs can be enumerated as follows:
  - PEVs produce less undesired emissions.
  - Electric motors produce almost no noise.
  - Electric motors require less regular maintenance.
  - Electric vehicles have higher efficient compared to the ICEV.
  - The propulsion system and the drive train in electric vehicles have simpler designs since BEVs do not require certain components such as transmission, clutch, spark plugs, and starter to name a few.
  - Even though BEVs do not have tailpipes producing electricity to fuel, they still cause emissions.

- Charging vehicles with clean generation mix (e.g., hydro, wind, and nuclear) leads to less emissions compared to fossil-based power generation options, such as coal and diesel.
- The United States is leading in gross PEV sales, while Norway has the highest percentage of PEVs in new vehicle sales.

References

[1] *Greatest Achievements of the 20th Century*, National Academy of Engineering, http://www.greatachievements.org/.

[2] Boulanger, Albert G., et al., "Vehicle electrification: status and issues", *Proceedings of the IEEE*, Vol. 99, No. 6 2011, pp. 1116-1138.

[3] Chan, C. C. "The Rise & Fall Of Electric Vehicles In 1828-1930: Lessons Learned", *Proceedings of the IEEE*, Vol. 101, No. 1, 2013, pp. 206-212.

[4] Hawkings, T., et al., "Comparative Environmental Life Cycle Assessment of Conventional and Electric Vehicles", *Journal of Industrial Ecology*, Vol. 17, issue 1, Feb. 2013, pp. 53-64.

[5] Wakefield, E., *History of the Electric Automobile*, SAE International, 1993.

[6] Chan, By CC. "The state of the art of electric, hybrid, and fuel cell vehicles." *Proceedings of the IEEE*, Vol. 95, No. 4, 2007, pp. 704-718.

[7] Husain, Iqbal. *Electric and Hybrid Vehicles: Design Fundamentals*, CRC Press, 2011.

[8] Larminie, James, and John Lowry, *Electric Vehicle Technology Explained*, John Wiley & Sons, 2004.

[9] Rajakaruna, Sumedha, et al., *Plug In Electric Vehicles in Smart Grids*, Springer Verlag, 2015.

[10] Garcia-Valle, Rodrigo, and Joao A. Pecas Lopes. *Electric Vehicle Integration into Modern Power Networks*. Vol. 2. Springer Science & Business Media, 2012.

[11] United States Environmental Protection Agency, *Greenhouse Gas (GHG) Emissions for Electric and Plug-in Hybrid Electric Vehicles*, http://www.fueleconomy.gov/feg/Find.do?action=bt2.

[12] Argonne National Laboratories, Energy Systems Division, *The Greenhouse Gases, Regulated Emissions, and Energy Use in Transportation Model*, greet.es.anl.gov.

[13] World Nuclear Association, *Comparison of Lifecycle Greenhouse Gas Emissions of Various Electricity Generation Sources*, 2011.

[14] United States Environmental Protection Agency, *Fuel Economy Guide* www.epa.gov/fueleconomy/data.htm.

Chapter 2

Enabling Technologies

Successful adoption of PEVs strongly hinges on their performance and their adequate versatility to address users' needs, expectations, and preferences. Analyzing the gap between the current enabling technologies for battery storage, vehicle powertrain, and electric vehicle charger equipment on the one hand, and the required performance targets on the other hand, is instrumental in promoting mainstream adoption of PEVs. This chapter presents an overview of existing technologies involved in the current PEVs.

2.1 Battery Technology

Battery storage is the key enabling technology for facilitating the mainstream adoption of PEVs, since it heavily influences several major measures pertinent to vehicle performance (e.g., the cost of adoption), the driving range, and reliability. Small-scale batteries have a mature technology, which adequately serve the needs of a broad range of electronic equipments that are designed to last for about three years. Nevertheless, such small-scale batteries lack the sufficient capacity to properly serve PEVs. Large-scale batteries, on the other hand, are the proper alternatives for PEVs, but deploying them faces several technological and operational challenges, which must be addressed prior to being deployed in PEVs. Specifically, large-scale PEVs arc primarily designed to be deployed at stationary locations for the purpose of assisting certain critical missions in power grids, such as emergency backup and renewable integration. Because they were developed for being stationary, deploying them for PEV applications posed new challenges, as these batteries are required to exhibit high degrees of mobility. Furthermore, these batteries are subject to strenuous conditions such as over-discharge, short circuit, and vibrations. Besides these challenges, the size and weight of large-scale batteries deployed in PEVs should also be optimized to satisfy certain physical requirements pertinent to the desirable PEV performance measures such as acceleration, top speed, braking distance, and driving range. This section is devoted to discussing the present electric vehicle battery technology in detail, and the associated challenges that need to be addressed.

2.1.1 Lithium-ion Batteries

As shown in Table 2.1, the lithium-ion (Li-ion) battery technology is the most commonly used one in the current market. In Li-ion batteries, electrical energy is generated through electrochemical oxidation-reduction reactions. The basic component of a Li-ion battery is a cell, and each cell has five basic components that facilitate the energy conversion process. The first component is an anode, which gives up the electrons to an external circuit (e.g., electric motor) and oxidizes during the reaction. The second component is a cathode, which accepts the electrons from the external circuit. The third component is the process in which electrolyte mediates ion transfer inside the cells. The fourth component is a membrane that acts a separator and prevents the internal short circuits between the anode and the cathode, and the last component is a container, which is devised for safety purposes.

Material science and technology play a pivotal role in the advancing battery technologies. In a typical battery, the positive electrode is made up of lithium and transition metal material, while graphite is widely employed as the negative electrode material. The commercially available electrochemical couples include graphite-LiMN2O4, graphite-LiMn2O4, and graphite-LiFePO4. During the charging stage, lithium ions move from the positive electrode, pass through the electrolyte, and intercalate between graphite layers. During the discharge process, the reverse process takes place and the stored energy is used to run the electric motor.

Moreover, there has been an increasing interest in developing new battery technologies that offer higher energy densities and lower costs. Recent research activities on that front have been focusing on studying new nanomaterials to develop lithium-sulfur batteries that can extend the all-electric range of PEVs by threefold [1]. This research domain, while being increasingly promising, is still in its infancy stages.

Table 2.1
Current PEV Battery Manufacturer

Company/Model	Year	Type	Battery Type	Supplier
BMW i3	2014	BEV	Li-ion	Samsung SDI
Cadillac ELR	2015	PHEV	Li-ion	LG Chem
Chevrolet Spark	2015	BEV	Li-ion	LG Chem
Chevrolet Volt	2015	PHEV	Li-ion	LG Chem
Ford Fusion Energi	2014	PHEV	Li-ion	Panasonic
Ford Focus	2015	BEV	Li-ion	LG Chem
Honda Fit	2014	BEV	Li-ion	Toshiba
Honda Accord	2014	PHEV	Li-ion	Blue Energy
Kia Soul	2015	BEV	Li-ion Polymer	SK Innovation
Nissan Leaf	2015	BEV	Li-Ion	AESC
Tesla Model S	2014	BEV	Li-ion	Panasonic
Volkswagen e-Golf	2015	BEV	Li-ion	VW-Panasonic

2.1.2 Basic Battery Terminology

There exist certain definitions and measures that quantify and characterize the performance of the batteries for automotive applications. In this section, we review some of the most widely used definitions, which will serve as a quick reference for the rest of the discussions in this book.

- *Cells, modules, and packs* are the central physical components of a PEV battery. A *cell* is the smallest possible battery that contains electrodes, a separator, and an electrolyte. The output voltage of a typical cell varies between one to six volts. A *module* consists of multiple cells that are usually connected in either series or parallel. Similarly, a *pack* is composed of a few modules located in a single location.

- *Ampere-hour (Ah) capacity* is the most popular rating metric that measures the capacity of a battery. Ampere-hour capacity is the amount of charge transferred by one-ampere flow during one hour. Similarly, the *C-rate* or the *nominal C-rate* of a battery represents the charge rate (in amperes) of a battery during one hour. For instance, the *C-rate* of a 2.5 Ah battery is 2.5 A. The *E-rate* is used to measure the amount of *power* required to deplete the battery in one hour. The *rated Ah*, on the other hand, is defined as the nominal capacity of a new battery under the operating conditions (e.g., temperature and C-rate), which are specified by the manufacturer. Another frequently used capacity term is the *rated watt-hour capacity*, which is the product of *rated Ah capacity* and *rated battery voltage*. Apparently, a higher battery capacity provides longer driving ranges, and it also takes more time to charge the battery.

- *State of charge* (SoC) is an indicator of battery degradation over time, which can be caused due to operating conditions. Accurate gauging is critical for battery management systems. SoC measures the ratio of the present battery capacity with respect to maximum capacity, and is calculated according to

$$\text{SoC} = \frac{\text{Present Amount of Charge}}{\text{Total Amount of Usable Charge}} \; . \tag{2.1}$$

- *Depth of discharge* (DoD) is used to represent the percentage of discharged capacity. Hence,

$$\text{DoD} = 1 - \text{SoC}. \tag{2.2}$$

  Deep-cycle batteries are usually discharged to 80% or higher.

- *State of health* (SoH) is used to represent the ratio of the maximum charge capacity to its ideal conditions. The SoH indicates the performance degradations of the battery, which is also related to its remaining lifetime.

- *Cycle life* is the number of charge-discharge cycles that a battery can accommodate within the desired performance regions. Cycle life is determined for a particular DoD, which is typically around 80%. It is noteworthy that cycle life is computed for nominal charge-discharge conditions. Therefore, actual life of the battery can be shorter due to undesired operating conditions such as high temperature and humidity.

- *Calendar life* is the number of calendar years that a battery is expected to last, and it depends on the SoC and the operating temperature.

- *Specific energy* represents the amount of energy that can be stored in a unit mass, and it is often measured with the physical dimension watt-hours per kilogram (Wh/kg). It is computed according to

$$\text{Specific Energy} = \frac{\text{Rated Ah Capacity} \times \text{Rated Battery Voltage}}{\text{Battery Mass in kg}} . \tag{2.3}$$

  Specific energy, which is also referred to as the gravimetric energy density, is an indicator of the battery chemistry and packaging, which determines the additional weight introduced by the battery for a desired all-electric-range. It is one of the most important metrics for assessing the PEV performance.

- *Specific power* is the metric used to define the maximum available power per unit mass, and it is often expressed in watts per kilogram (W/kg). Specific power, sometimes also referred to as gravimetric power density, can be calculated via

$$\text{Specific Power} = \frac{\text{Rated Peak power}}{\text{Battery Mass in kg}} . \tag{2.4}$$

  Specific power determines the battery weight for particular PEV performance measures such as acceleration and regenerative braking.

- *Energy density* of a battery is the nominal energy per unit volume, and it is represented by watt-hour per liter (Wh/L). Energy density, or volumetric energy density, determines the physical space required to achieve a particular performance target such as driving range. Similarly, *power density* is the maximum available power per unit volume of the battery, and it is represented by watt per liter (W/L).

- *Maximum continuous discharge current* is the peak current at which the battery is depleted uninterruptedly. This parameter also serves as a threshold that prevents excessive discharge rates, and therefore, protects the battery life. Moreover, maximum continuous discharge current, along with the specifications of the electric motor, determine the top speed and the acceleration of the PEV.

- *Maximum 30-second discharge pulse current* is the maximum current at which the battery can be discharged with pulses up to 30 seconds. This limit is determined by the manufacturer to protect the battery's lifetime. Similar to the previous performance measure, it affects the acceleration performance of the vehicle.

- *Float voltage* is the steady state voltage level at which the battery is charged after reaching 100% SoC in order to compensate the self-discharge losses.

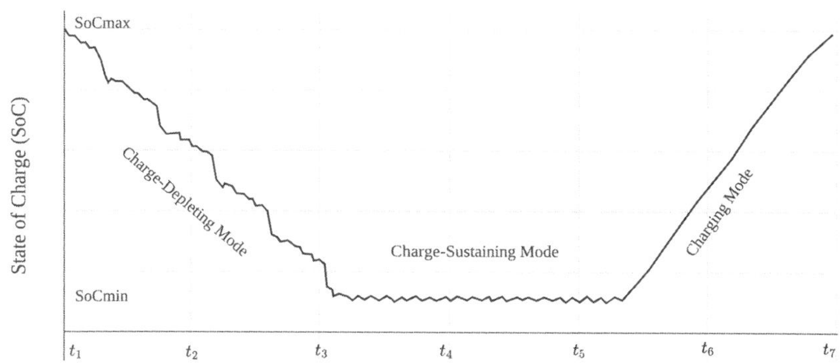

Figure 2.1 PEV operational modes: charge-depleting, charge-sustaining, and charging.

- *Cut-off voltage* is defined as the minimum allowable voltage for the battery, which describes the empty state of the battery. It is determined by the manufacturer.

- *Internal resistance* is simply the equivalent resistance of the battery, which is different for charging and discharging operations.

- *Charge sustaining* (CS) mode refers to the SoC levels, typically ranging between 25% and 30%, at which a vehicle exhausts its all-electric range and starts being fueled only by only gas. Hence, this mode exists only in PHEVs.

- *Charge depleting* (CD) mode refers the SoC range of a PEV where the vehicle is used to generate the traction power. CD is composed of two submodes: the *blended mode*, which only exists for PHEVs and is the stage at which the PHEVs operate at low speed, and the *all electric mode*, in which only the stored energy is consumed. For BEVs, this mode corresponds to 20%-80% SoC, while for PHEVs the SoC range is 30%-100%. CS and CD modes are shown in Figure 2.1.

Table 2.2: USABC Advance Battery Goals for PEVs in 2020

End of Life Characteristics (30°C)	**Unit**	**Target**	**Corresponding Vehicle Performance**
Peak discharge power density, 30 s pulse	W/L	1000	Acceleration and size
Peak specific discharge power, 30 s pulse W/kg	W/kg	470	Acceleration and weight
Usable energy density at $C/3$ discharge rate	Wh/L	500	All-electric-range and size
Usable specific energy at $C/3$ discharge rate	Wh/kg	235	All-electric-range and weight
Usable energy at $C/3$ discharge rate	kWh	45	All-electric-range, vehicle cost
Calendar life	Years	15	Life-cycle cost
Cycle life	Cycles	1000	Life-cycle cost
Selling price for 100k units	$/kWh	125	Acquisition cost
Operating environment	°C	-30°C to 52°C	Battery life
Slow charging time	Hours	<7 with J1772	Battery size cost
Fast charging	Minutes	15 min to 80 SoC	Battery size and cost
Maximum operating voltage	Volts	420	Battery health
Minimum operating voltage	Volts	220	Battery health
Peak current, 30 s	Ampere	400	Acceleration
Maximum self-discharge	% per month	<1	All-electric-range

2.1.3 Design Criteria

If not designed properly to meet the necessary performance requirements, the battery technology becomes a bottleneck in the early adoption phase of the PEVs. To establish a uniform set of standards for the industry, the United States Advanced Battery Consortium LLC (USABC) [1] has put forth a set of performance objectives for batteries, which are listed in Table 2.2. Based on these objectives, the following critical performance parameters should be carefully considered in PEV applications:

1. *Lifespan* of a battery is an important performance criterion to compete with gas-powered vehicles, as the cost of the batteries is still high. As explained above, there are two primary metrics for measuring the lifespan of batteries, namely cycle life and calendar life. Based on these metrics, the USABC, for instance, aims to have the technology of mass-producing pure electric vehicles of 15-year calendar life and a 1000-cycle life.
2. *Vehicle performance* should be comparable or superior to that of gas-powered vehicles to secure gain customer acceptance. PEVs deployed in regions with severe weather conditions (extremely hot or cold) will face additional challenges, as batteries are typically designed to work properly in mild weather conditions. The impact of temperature on battery lifetimes is well-documented in [2]. This document concludes that battery lifetimes are critically sensitive to temperature. Furthermore, solar loading can also lead to further battery heating and reduces battery life. Hence, the USABC is aiming to develop a technology to manufacture batteries that can operate properly in the range of -30°C to 52°C without experiencing major degradations.
3. *Specific energy and power* characteristics of the batteries affect various performance metrics, such as the driving range and weight of the vehicle. Current battery cells have nominal energy densities in the range $140 - 180$ (Wh/kg), whereas gasoline has the energy density of $13,000$ (Wh/kg). The USABC is aiming to improve the specific energy characteristics from the current levels to 235 (Wh/kg). On the other hand, current battery technologies are mature in terms of specific power as PEV batteries can discharge in a short time during acceleration of the vehicle. Therefore, the current specific power levels are marginally sufficient.

1 USABC is comprised of three industrial partners, General Motors, Ford, and Chrysler, as well as several national laboratories.

4. *Safety* has the utmost significance in PEV batteries, as the vehicles might operate in harsh environmental terrains and conditions, which include exposure to extreme temperatures, severe vibrations, and high rates of charge and discharge. Moreover, due to entropy changes Li-ion batteries can ignite if overcharged. The United States National Renewable Energy Laboratory (NREL) specifies the guidelines for battery safety in order to reduce the likelihood of battery failures and remedies to lessen the outcome of incidents such as fires and explosions [3].

2.1.4 Battery Management Systems

A battery management system (BMS) links the battery to the vehicle in order to maximize the vehicle's performance by protecting the batteries' health while minimizing power consumption. To achieve these objectives, the BMS should monitor, control, and communicate with the battery. Therefore, the primary functions of a BMS are:

- Under voltage and over voltage protections;
- Short circuit protection;
- Thermal protection;
- Cell balancing;
- Controlling battery charging and discharging;
- Determining battery SoC and SoH;
- Safety protection.

Figure 2.2 demonstrates a block diagram of a generic BMS and the dynamics of its major functions. In order to perform the functions listed, the sensing unit gathers certain pieces of information such as voltage and temperature from the battery, and communicates them to the management and the balancing units. The role of the balancing unit is to ensure that cell voltages operate within the allowed limits. Voltage, current, and temperature readings are used to determine the SoC levels of the batteries,. Determining SoC levels is a challenging task since other factors such as battery age and manufacturing conditions also affect the accuracy of the SoC measurements. There exist several techniques to enhance the accuracy of SoC measurements, and some of the most effective ones are current integration, impedance spectroscopy, direct current resistance, and Kalman filtering.

Cell balancing is an important part of battery management systems. In a typical PEV battery, several cells are connected in parallel and act as a block to meet the

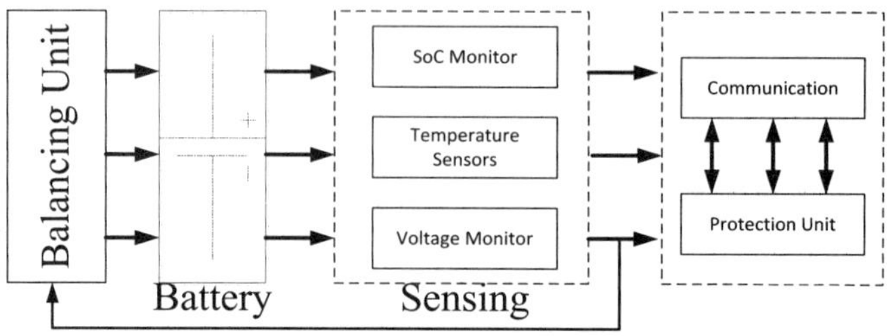

Figure 2.2 Generic battery management system block diagram.

high capacity requirements, while several cells are wired in series to meet high voltage requirements. Moreover, each cell can have different properties due to chemical offsets. Therefore, cells may have different voltage levels, which may degrade batteries' lifetimes or even cause life-threatening incidents. Cell balancing is required to balance cell voltages, and subsequently, prevent such unwanted consequences. Some of the widely used techniques include deploying dissipative resistors, buck-boost shunting, multi-winding transformers, and switched capacitors.

2.2 Plug-in Electric Vehicle Drivetrain Architectures

The terms drivetrain and powertrain are used to define the combination of the transmission and the propulsion units. In BEVs, the powertrain has a simpler design than its counterpart in ICEVs since it does not involve a clutch, and in some architectures not a gearbox. As explained in the previous chapter, BEVs are entirely propelled by an electric motor running on the battery-stored electricity. PHEVs, on the other hand, can be powered by either an electric motor or an internal combustion engine. Therefore, the architecture of PHEVs is more complex than that of BEVs. The powertrain design for BEVs is simpler as the traction power is only converted from the stored electricity, and BEVs do not employ combustion engine and transmission systems. For PHEVs, there are two main powertrain architectures: series and series-parallel architectures. An overview of the diagrams of these architectures is presented in Figure 2.3.

In the series PHEV architecture, the vehicle is propelled entirely by the electricity and the energy coming from both the battery and the combustion engine is combined in a direct current (DC) bus. Therefore, the combustion engine has no

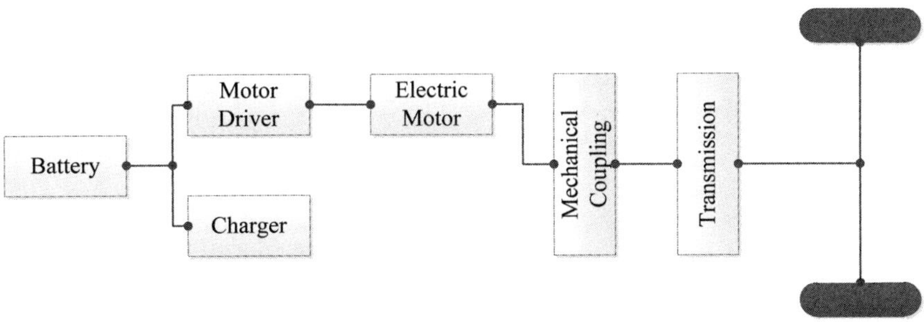

(a) BEV architecture.

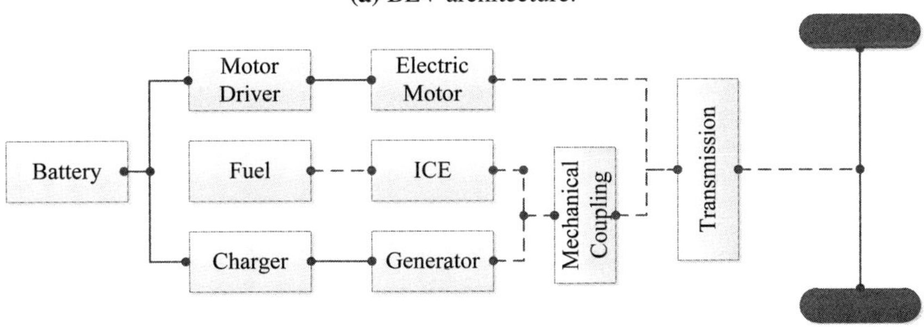

(b) PHEV series architecture.

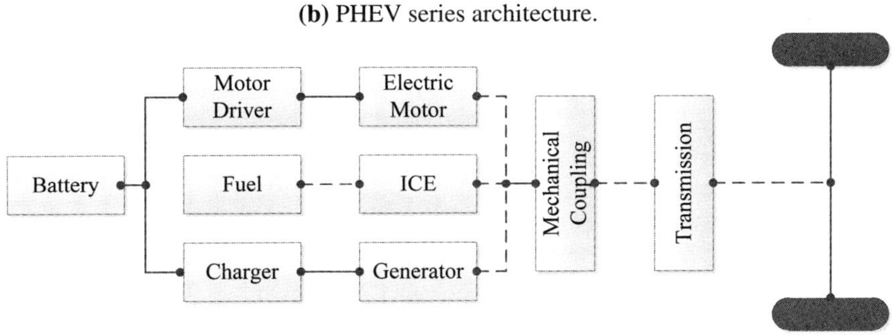

(c) PHEV series-parallel architecture.

Figure 2.3 Powertrain architectures.

connection with the traction load and does not directly move the vehicle. In series-parallel architecture, however, both mechanical and electrical power couplers are employed. Therefore, the vehicle can be run either by the electric motor or by the combustion engine. As expected, and demonstrated in Figure 2.3, this architecture is more complicated and costly.

PHEV models can also be categorized by their levels of hybridization. Hybridization factor (HF) is a metric used for this purpose and represents the relative value of the power of the electric motor to the total power of the vehicle. Therefore, it can be computed according to

$$\mathrm{HF} = \frac{P_{\mathrm{EM}}}{P_{\mathrm{EM}} + P_{\mathrm{ICE}}}, \tag{2.5}$$

where, P_{EM} represents the peak power of the electric motor and P_{ICE} is the peak power of the internal combustion engine. For a conventional ICEV we have $\mathrm{HF} = 0$, while the HF of a BEV is 1, and the hybridization factor of PHEVs ranges between zero and one.

2.3 Electric Vehicle Service Equipment

Electric vehicle service equipment (EVSE) refers to the infrastructure required to charge the battery of a PEV. The deployment of EVSEs, which are also commonly known as charging stations, is critical for accelerating PEV penetration since they are required to overcome range anxiety. The physical components of an EVSE include a control box, an electric vehicle connector, and power cords. EVSEs are typically connected to the power grids to provide energy transfer to the vehicle. The capacities of EVSEs can vary significantly, which in turn determine the charging duration when they are aligned with the ratings of the onboard chargers.

An electric vehicle connector physically connects the vehicle with the EVSE. The type of the connector determines the compatibility of vehicles with the charging stations. For instance, Tesla uses Supercharger connectors, Society of Automative Engineers has developed the SAE J1772, and Japanese standard CHAdeMO connectors are used in fast charging stations. On the PEV side, there are two important components that are related to the charging equipment. The first one is the PEV inlet, which allows the connector to be plugged in. Some PEV models are equipped with more than one inlet. Onboard battery charger converts alternative current (AC) power into direct current (DC) power so that it can be stored in the battery. On the

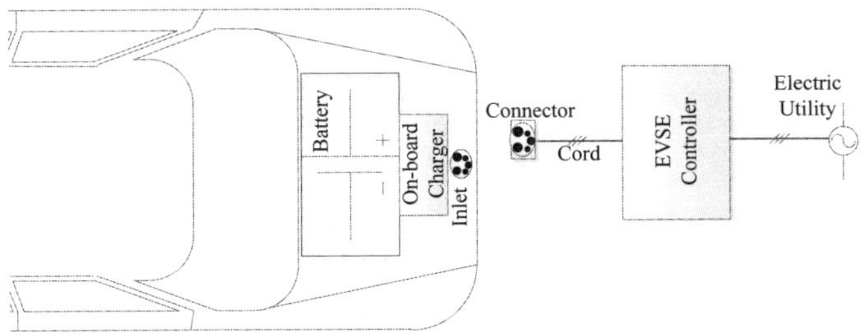

Figure 2.4 Plug-in electric vehicle charging infrastructure.

other hand, if DC fast chargers are used, the battery charger is bypassed and the EVSE directly fills up the battery. An overview of EVSEs is provided in Figure 2.4. Standardizing PEV charging equipment is an indispensable process, necessary for enhancing service convenience through eliminating the need for a multiplicity of cords and adapters for long-distance trips. Next, we provide a detailed analysis of charger technology, charging standards, and the associated infrastructures.

2.3.1 Battery Charger Technology

Electric chargers replenish PEV batteries by converting alternating current to regulated direct current. As we briefly discussed in the previous section, battery charging technologies are critical elements as they directly influence charging durations and affects batteries' life cycles. Overall, battery charging technologies can be categorized, broadly, into *onboard* and *off-board* charging. Both charging technologies can be further divided into two groups, where one allows only *unidirectional* power flow, and the other one enables *bidirectional* power flow.

As their name suggests, onboard chargers are physically located in the PEV. Onboard chargers are used primarily for low charging power levels, which are typically less than 10 kW. Therefore, they do not cause battery heating and, consequently, they protect batteries' health. Also, such chargers are used on users' premises or public parking lots where there is not enough physical infrastructure for off-board chargers. Onboard chargers, however, could become bottlenecks in operating PEVs due to their weight, cost, and physical space requirements. On the other hand, off-board chargers are designed for higher charging rates (i.e., 10 − 80 kW) and therefore require more elaborate power electronics architectures. In order to facilitate high levels of charging power, additional infrastructure, such as

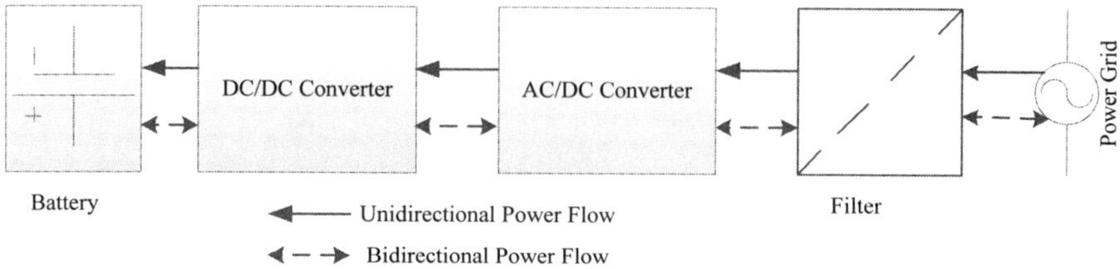

Figure 2.5 Unidirectional and bidirectional converter technology.

energy storage units or three-phase lines, might become necessary. Most public DC charging stations deploy off-board chargers, and since they are not located on the vehicles, they are less constrained by their physical sizes and weights. The major setback of off-board chargers is that they degrade battery lifetime more quickly than the onboard ones, due to involving high-voltage energy transfer.

Both onboard and off-board chargers can provide unidirectional or bidirectional power flows. Typically, unidirectional chargers have simpler hardware architecture and only allow the grid to charge the battery. The operations of unidirectional chargers are easier to control, especially when high PEV penetration impacts the feeder operations. Usually, unidirectional chargers employ a diode bridge, a filter, and DC-DC converters. A generic topology of these charges is presented in Figure 2.5. Bidirectional chargers, in contrast, enable the vehicle to inject power back to the grid, which allows the PEVs to act as mobile storage units. Such a bidirectional flow, as a result, provides ancillary services, such as voltage control, frequency control, and emergency backup, and assists with the operations of the power grid. Bidirectional power flow, however, degrades battery lifetime significantly since it involves frequent charging and discharging cycles. The combined effects of acting as an ancillary service (advantage) and degrading battery quality (disadvantage) determine whether deploying bidirectional charging outperforms the unidirectional counterpart. For instance, the study in [4] examines the economic benefits of PHEV batteries for grid storage, where it shows that the annual benefits of bidirectional charging[2] can be in the range of $140-250$ USD when using perfect batteries. Nevertheless, by accounting for the degradation cost, the effective amount of saving reduces to $10-120$ USD, where the exact amount depends on the tariffs.

2 Study conducted in 2009, assuming A123 Systems LiFePO4/Graphite cells, 16 kWh battery pack, and a replacement cost of 5000 USD.

Naturally, such benefits increase as the cost of storage drops. Besides such cost-benefit analyses, bidirectional charger units are also more costly than unidirectional ones. These setbacks, in conjunction with some other technology-related necessities, make bidirectional charging less attractive in the existing technologies. For instance, chargers regulate the power quality by drawing low distortion current and high levels of power factor to maximize the available real power [5]. Hence, to allow for bidirectional power flows, battery chargers should comply with standards, such as IEEE 1547, SAE-J2894, and IEC 1000-3-2, which regulate the allowable harmonics and DC injections into the grid.

2.3.2 Charging Rates and Options

In order to promote mainstream acceptance of electric vehicles, *availability* of charging stations that provide charging service at various locations and rates to meet customer needs is critical. Such facilities are required to beat the range anxiety resulting from the current limited ability to recharge them in a time commensurable with filling the tank of a gas-powered ICEV. Even though range anxiety can be partly addressed by exploiting driver behavior (e.g., charging vehicles during the night (garage charging) when there is spare capacity in the grid) such charging options are not necessarily available to all PEV owners (e.g., those living in dense metropolitan areas). Hence, the need for longer driving ranges requires the deployment of public charging facilities. During weekdays, PEV owners can spend several hours parked at various locations (e.g., workplaces, schools, and shopping malls). Hence, charging service at parking lots could provide the necessary fuel to extend the driving range.

The charging options mentioned above are classified by the rate at which the batteries are charged. For instance, one hour AC level 1 charging can provide 2 to 5 miles of driving range, while level 2 can offer 10 to 20 electric miles with the same duration. DC fast chargers typically use high charging power, which can replenish $50 - 70$ miles within 20 minutes. As the examples suggest, AC level 1 is designed to provide basic service, for example for home or garage charging with standard home outlets. AC level 2 charging, on the other hand, uses higher power and the chargers are typically located at big complexes such as medical centers, universities, airports, shopping malls, and industrial parks. *Park and charge* and *shop and charge* are among the common terms used to describe this option. DC fast charging stations are designed to compete with gas stations and provide charging services in short durations. Level 2 and DC fast stations are also considered as public charging facilitics and are primarily designed to serve customers living in urban and

Table 2.3
Charge Methods Electrical Ratings (Europe)

Method	**Connection**	**Power (kW)**	**Max. Current (Amps)**
AC level 1	1-phase	3.7kW	10 − 16A
AC level 2	1 or 3-phase	3.7 − 22kW	16 − 32A
AC level 3	3-phase	22kW+	32A+
DC fast charging	DC	22kW+	3225A+

multifamily environments where there is no access to garage charging. Apart from the charging rates, level 2 stations serve a large pool of customers, while DC fast stations are small scale due to the amount of the power they can provide [6, 7].

Charging rates and durations vary in different countries primarily due to their different power system standards. In the next few sections we provide an overview of some of the major standards developed to be adopted in different regions.

2.3.2.1 European Charging Standards IEC 61851

The European standards are determined by the International Electrotechnical Commission (IEC). The technical committee IEC TC69 along with its associated workgroups characterize the IEC 61851 standards. The primary goal of this task force is to develop internationally acceptable standards for compatibility, safety, and performance. From a compatibility standpoint, deploying the same charging plugs and sockets enables PEVs to have access to service everywhere, while the safety aspects aim to reduce the risks associated with PEV charging, especially at the street level where non-specialist users will be using the charging equipment. The performance aspect is concerned with optimizing and regulating energy flow between the grid and the PEVs. The IEC charging standards include

- IEC 61851-1:2010 defines the electric vehicle conductive charging system, for both onboard and off-board charging up to AC voltage of 1000 V and DC voltage of 1500 V.
- IEC 61851-21:2001 determines the electric vehicle requirements for conductive connection to AC/DC supplies.

Table 2.4
Charging Methods (North America)

Method	Nominal Supply	Max. Current	Branch Circuit Breaker Rating
AC level 1	120 vac, 1-phase	12A	15A (minimum)
AC level 2	208 – 240 vac, 1-phase	32A	40A
DC fast charging	600 vdc max.	400A maximum	As required

- IEC 61851-22:2001 provides the AC charging requirements equipped with supply voltages with outputs up to 690 V.
- IEC 61851-23:2014 defines the requirements for DC charging up to 1500 V DC.
- IEC 61851-24:2014 establishes the requirements for digital communication between DC charging station and the PEV to control the power flow.

Moreover, the DC fast charging has been recognized by the IEC as an international DC charging standard. The IEC 61851-23 standard defines the requirements for DC chargers and presents communication requirements for the charger and the PEVs. Apart from CHAdeMO, GB/T by China, COMBO1 by the United States, and COMBO by Germany have been recognized by IEC as fast charging standards.

In IEC 61851 standard, IEC defined four charging modes: mode 1 charging for non dedicated outlets, mode 2 charging for non dedicated outlets with in-cable protection devices, mode 3 for dedicated outlets, and mode 4 for DC connections. Moreover, this document also defines the charging power levels. An overview of the European charging standards is presented in Table 2.3.

2.3.2.2 North American Charging Standards SAE J1772

In the United States, the Society of Automotive Engineers (SAE) has been developing the J1772 standard for electric vehicle charging that covers physical, electrical, communication protocol, and performance requirements for PEV charging. Similar to the IEC standards, the SAE defines three charge levels: level 1, level 2, and DC fast charging (a detailed overview is provided in Table 2.4). Level 1 charging contains a standard three-prong plug on one side and a connector at the other side.

J1772 is the first PEV charging standard that has gained consensus in the industry. This standard defines the electrical and physical characteristics of the vehicle inlet and the connector. Apart from the SAE fast charging standard, Tesla has also developed Tesla Supercharger to be used exclusively by Tesla vehicles. Hence, in the United States there exist three competing DC fast charging standards: ChAdeMO, SAE COMBO, and Tesla SuperCharger. Even with discrepancies, there exist charging technologies that comply with all three standards so that different vehicle models can use a public station. Moreover, the SAE defines SAE J2847-1 standards for the communication between PEVs and the power grid, and SAE J2836-1 for communication between PEVs and the utility. The details of the two standards are given in Table 2.5.

Table 2.5

Overview of Charging Standards in the United States

Standard	Note
SAE J1772	PEV conductive charge system and coupler general requirements in the North America.
SAE J2836/1	Defines the communication uses cases required for energy transfer between PEVs and the power grid.
SAE J2836/2	Defines the communication uses cases required for energy transfer between PEVs and the EVSE.
SAE J2836/3	Defines the communication uses cases needed for reverse energy transfer between PEVs and the power grid.
SAE J2836/4	Defines the use cases for the communication for PEV diagnostics.
SAE J2836/5	Defines the use cases between PEVs and customers.
SAE J2836/6	Defines wireless charging use cases.
SAE J2847/1	Defines the communication between PEVs and the utility grid.
SAE J2847/2	Identifies messages for DC energy transfer from the grid to the PEV.
SAE J2847/3	Identifies messages for DC energy transfer from the PEV back to the grid.
SAE J2847/4	Defines communication requirements for diagnostics between the PEV and the EVSE.
SAE J2847/5	Defines communication requirements between PEVs and customers.
SAE J2847/6	Defines communications for wireless charging.

2.3.3 Wireless Charging

Wireless charging, or inductive power transfer, is an emerging technology and enables PEVs to charge their batteries in a contactless manner. Wireless charging has recently gained some attention mainly because it can allow power transfer between the grid and a vehicle while the vehicle is on the move. Based on this charging technology, all electric range can be extended without requiring the vehicle to spend time for the charging service and the required battery size can be reduced to [11,12]. Other benefits of wireless charging are that it requires limited interaction with PEV owners and eliminates the hazards associated with cords and electric sparks.

Nevertheless, wireless charging has several disadvantages. The most significant concern is safety. Specifically, PEV charging requires high-frequency magnetic fields between transmitting and receiving coils. Typically the transmitting coils are built into a parking space and the receiving coils are attached to the underbody of the PEVs. Depending on the air gap between the coils, the flux leakage might have health risks for an individual near the coils. Moreover, the efficiency of wireless chargers are lower than the wired chargers and the charging rates are typically slower than the wired ones. For instance, recent tests show that charging power can reach up to 3.6 kW when the distance between a vehicle and a charging coil is 0.2 meters [13]. Moreover, higher costs compared to conductive chargers and sensitivity to vehicle alignment are additional disadvantages. The frequency of charging is also a critical parameter. The studies in [14] show that the efficiency of the power transfer is subject to severe degradations when a fixed frequency is used and the vehicle positioning is imperfect. To overcome this issue, transmit frequency is adjusted according to the position of the two coils. Since 2012, several companies, including Qualcomm [15], WiTricity [16], and Plugless Power [17], have introduced commercial products for wireless PEV charging.

2.3.4 Charging Station Deployment

The global PEV charging equipment market is expected to grow from more than one million units in 2014 to more than twelve million units in 2020. The current portfolio of deploying chargers is broadly divided into residential and commercial segments. Residential charging deployment, which can either include level 1 or level 2 chargers, is higher than that of commercial stations largely because residential charging is preferred by many early PEV adopters. Moreover, PEV owners enjoy decreased prices resulting from increased commercialization of charging

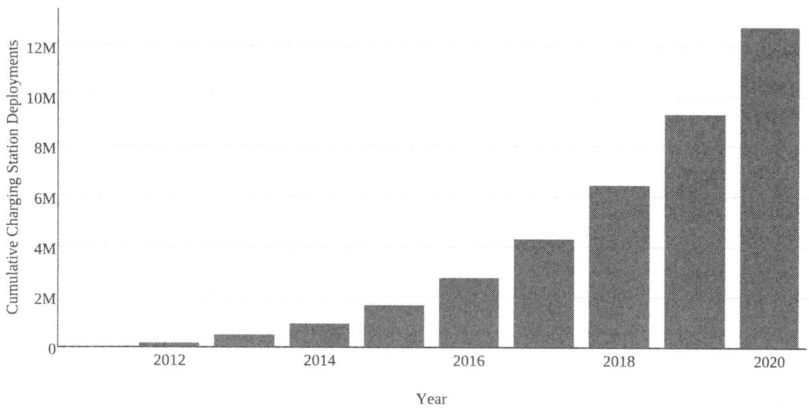

Figure 2.6 Charging station deployments. (Source: IHS Automative).

products. The technologies supporting commercial charging stations have also been advancing over the past few years, and currently there exist a variety of commercial hardware and software solutions for public charging stations.

In Figure 2.6, we present global charging station deployment trends. Clearly, increasing PEV sales trigger charging station deployments. For instance, in Japan the number of charging stations has already passed the number of gas stations. Similar to PEV sales, the United States is leading the EVSE market. As of October 2015 there exist more than $27,000$ charging stations dispersed around more than $10,000$ locations. An overview of charging stations across the United States is depicted in Figure 2.7.

2.4 Chapter Summary

This chapter has focused on enabling technologies for PEVs. The key technologies discussed in this chapter involved battery technology, PEV drivetrain architectures, electric vehicle service equipments, and associated industrial standards. A major portion of the battery technology has been devoted to lithium-ion batteries as they are the most widely used type in the current vehicle market. A comprehensive

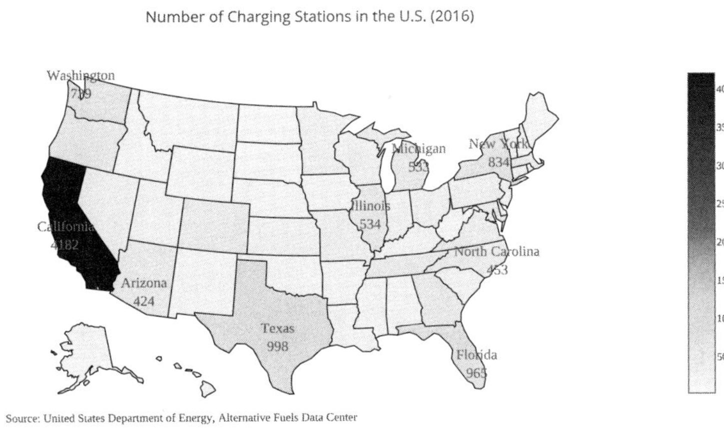

Figure 2.7 Number of charging station by state in the United States, December 2016 [10].

examination of battery terminology and the related terminology were put forth to provide more insights into the complicated nature of storage devices. Battery management systems (BMS) are an integral component of PEVs and the basic BMS functionalities that aim to optimize battery performance, prolong battery life, and ensure battery safety were treated. We also presented different PEV drivetrain with various hybridization factors for BEVs and PHEVs. The last part of the chapter was allotted for electric vehicle supply equipment, converter technology, charger technologies, and associated standards. Charger types differ across the globe according to charge rate and connection types and the major standards are created in the United States, Europe, and Japan.

References

[1] Liang, X., et al., "A highly efficient polysulfide mediator for lithium-sulfur batteries," *Nature Communications*, Vol. 6, No. 5682, Jan. 2015.

[2] Pesaran, A., Smith, K., and Marke, T.,"Impact of the 3Cs of batteries on PHEV value proposition: Cost, calendar life, and cycle life," *Proc. Advanced Automotive Battery and EC Capacitor Conference*, Long Beach, CA, June 10-12, 2009, pp. 8-12.

[3] Doughty, D. and Pesaran, A., *Vehicle Battery Safety Roadmap Guidance*, National Renewable Energy Laboratory, Oct 2012.

[4] Peterson, S., Whitacre J.F., Apt, J., "The economics of using plug-in hybrid electric vehicle battery packs for grid storage," *Journal of Power Sources*, Vol. 195, No. 8, 2010, pp. 2377-2384.

[5] Yilmaz, M. and Krein, P. "Review of battery charger topologies, charging power levels, and infrastructure for plug-in electric and hybrid vehicles," *IEEE Transactions on Power Electronics*, Vol. 28, No. 5, 2012, pp. 2151-2169.

[6] Bayram, I.S., et al., F. "Local energy storage sizing in plug-in Hybrid electric vehicle charging stations under blocking probability constraints," *Proc. IEEE International Conference on Smart Grid Communications*, Brussels, Belgium, Oct., 2011, pp. 78-83.

[7] Bayram, I.S., et al., "Strategies for competing energy storage technologies for in DC fast charging stations," *Proc. IEEE International Conference on Smart Grid Communications*, Tainan City, Taiwan, Oct., 2012, pp. 1-6.

[8] Bayram, I.S., Michailidis, G., and Devetsikiotis, M., "Electric power resource provisioning for large scale public EV charging facilities," *Proc. IEEE International Conference on Smart Grid Communications*, Vancouver, Canada, Nov., 2013, pp.133-138.

[9] Bayram, I.S., et al., "Electric power allocation in a network of fast charging stations," *IEEE Journal on Selected Areas in Communications*, Vol. 31, No. 7, 2013, pp.1235-1246,

[10] www.afdc.energy.gov.

[11] Lukic, S. and Pantic Z. "Cutting the cord: Static and dynamic inductive wireless charging of electric vehicles." *IEEE Electrification Magazine*, Vol. 1, No. 1, 2013, pp. 57-64.

[12] Li, S. and Chunting, C. "Wireless power transfer for electric vehicle applications," *IEEE Journal of Emerging and Selected Topics in Power Electronics*, Vol. 3, No. 1, 2015, pp. 4-17.

[13] https://www.fraunhofer.de/en/press/research-news/2015/august/Wireless-charging-and-discharging-for-electric-vehicles.html.

[14] Krishnan, S. et al., "Frequency agile resonance-based wireless charging system for electric vehicles," *Proc. IEEE International Electric Vehicle Conference*, Greenville, SC, Mar., 2012, pp. 1-4.

[15] https://www.qualcomm.com/products/halo.

[16] http://witricity.com/applications/automotive/.

[17] https://www.pluglesspower.com/.

Chapter 3

Roadblocks to PEV Integration

The mainstream adoption of PEVs faces several challenges, since the incumbent ICEV technology is supported by a century-long production and design legacy. As a result, deployment of the ICEVs is more compatible with the well-developed service networks, and consequently, it is more convenient for the customers to continue relying on them. This influences the penetration of the PEVs, as it is ultimately linked to the customer perceptions of this technology. According to the United States National Research Council [1], as far as customers' acceptance is concerned, the barriers to the widespread PEV adoption are (1) the limited variety of PEV models; (2) range anxiety; (3) lack of sufficient information on electricity consumption, fuel cost comparison, and charging infrastructures; and (4) lack of enough information about environmental benefits.

In additional to all such barriers, the purchasing powers of households also is a defining factor, driven by the tendency of customers to make conservative decisions and choose options that bear fewer uncertainties, and at the same time, are affordable, reliable, and safe. Behavioral economists have conducted extensive studies on customer acceptance of new products, and it is well-documented that the PEV technology must provide enough benefits to offset the difference between the cost of the incumbent technology (ICEVs) and the risks and uncertainties involved in its electric counterpart [2]. The theory of *diffusion of innovations* analyzes market penetration rates of various new products [1], as well as the factors affecting customers' decisions. Based on such analyses, for even less costly and more typical products, such as consumer electronics and appliances, gaining mainstream

1 Details of diffusion of innovation models are presented in Chapter 13.

acceptance (above 50% penetration) takes more than a decade. In the case of PEV, it is expected that such mainstream acceptance will take several decades. Furthermore, according to [3], there exist three additional technology-specific roadblocks for PEV acceptance, namely: (1) battery cost; (2) availability of the facilities for roadside assistance and maintenance services; and (3) sufficient expansion of the charging infrastructures.

Even though PEVs have started to represent a sizable portion of national fleets, their penetration level falls behind the projections made in 2010. While barriers mentioned above are driven, predominantly, by economic and social challenges, there still exist technological challenges yet to be addressed. Among all such challenges, the most prominent ones are (1) the impacts of the PEV load on the power grids; (2) the design and operation of the charging infrastructure; and (3) the cost of ownership. The rest of this chapter is focused on providing detailed analysis on these challenges.

3.1 Impacts of PEV Charging

One of the major bottlenecks of PEV integration is the *readiness* of the power grid for accommodating the extra load they introduce. This is especially challenging as the current power grids, for the most part, were built decades ago without being modernized at the pace that energy consumption has grown. Moreover, upgrading the grid is a capital-intensive process; therefore, PEV demand should be carefully managed and aligned with the available resources.

3.1.1 Distribution Network

PEV charging and the associated infrastructure requirements affect various operations in the power grid. These impacts will be most strenuous on the operations in the distribution network. Some of these impacts imposed on the distribution network are (1) thermal loading; (2) harmonic distortions; (3) distribution system losses; and (4) system unbalances. In this section, we define each of these network characteristics and discuss the potential impacts of PEV charging on them.

3.1.1.1 Thermal Loading

Distribution transformers have thermal operating limits, which are necessary in order to enhance their security and lifetime. Penetration of PEVs is likely to take place in several stages due to a diverse range of preferences, financial incentives, and lifestyles of the adopters. Based on these factors, adopters can be clustered into distinct groups, each finding PEVs attractive at a particular stage of the evolution of the technology. PEVs emerging in clusters can cause thermal loading on the distribution grid. The transformer *loss-of-life*, or the aging acceleration factor, is governed by the load, the ambient temperature, and specific transformer temperatures. Spatiotemporal spikes in PEV demand can quickly increase the load and the operating temperature of the local transformer. To quantify the impacts of PEV demands on the transformers, we define the operating temperature of a transformer, also called the hottest-spot temperature, as

$$T_{\mathrm{HS}} \triangleq T_{\mathrm{A}} + \Delta T_{\mathrm{TO}} + \Delta T_{\mathrm{G}} \ , \tag{3.1}$$

where T_{HS} is the winding hottest-spot temperature, T_{A} is the average ambient temperature, ΔT_{TO} is the top-oil rise over the ambient temperature T_{A}, and ΔT_{G} is the winding hottest-spot temperature over the top-oil temperature. In order to compute the *loss-of-life*, we define the aging acceleration factor F_{AA} (for each 15 minutes) according to

$$F_{\mathrm{AA}} \triangleq \exp\left(\frac{15000}{383} - \frac{15000}{T_{\mathrm{HS}+273}}\right) , \tag{3.2}$$

where the constants are determined by the *IEEE Guide for Loading Mineral-Oil-Immersed Transformers* [4]. Accordingly, the equivalent aging factor for each month can be computed by

$$F_{\mathrm{EQ}} \triangleq \frac{\sum\limits_{n=1}^{N} F_{\mathrm{AA},n} \cdot \Delta t_n}{\sum\limits_{n=1}^{N} \Delta t_n} , \tag{3.3}$$

where n is the index of the time intervals, δt_n is the duration of the $n\mathrm{th}$ interval, $F_{\mathrm{AA},n}$ is the aging acceleration factor for the temperature during the $n\mathrm{th}$ time interval, and N is the total number of time intervals. Hence, the aggregate lifetime lost, in terms of hours, can be computed by properly normalizing F_{EQ} [2] through multiplying it by t, described in hours, and then dividing it by the normal insulation

2 Typically t is chosen as 24 hours.

life[3]; that is

$$\% \text{ of Loss of life} = \frac{F_{\text{EQ}} \times t}{\text{Normal insulation life}} \times 100\,. \tag{3.4}$$

Based on this definition, the fraction of the lost lifetime of a transformer operating at 110 Celsius is 0.0133, based on 180,000 hours of normal life as prescribed by the IEEE [4]. Based on this notion of loss-of-life, concurrent PEV charging imposes additional stress on the transformers, the effects of which can be quantified by the loss-of-life factor. Therefore, this provides a guideline for designing the charging strategies so that the physical constraints of the transformers are satisfied. Furthermore, geographical location and climate can also play an influential role in determining the lifetime of the transformers. Specifically, charging vehicles at high ambient temperatures (e.g., hot summer afternoons) has more adverse effects than charging the vehicle during the winter. As a case study, the analysis in [5] compares the impacts of PEV charging on a 25-kVA overhead distribution transformer serving 12 houses with 6 PEVs. This analysis shows that moderate levels of PEV demands (e.g., 50% penetration), may not have a severe impact on the operations of a city with the size of Burlington, VT, where the ambient temperature is low. However, if the same amount of load is applied to a transformer located in a warmer location, such as Phoenix, AZ, the transformer's lifetime reduces very rapidly.

3.1.1.2 Harmonic Distortion

In power systems, every nonlinear load, such as computers, uninterruptible power supplies (UPSs), fluorescent lighting, and PEVs, cause harmonic distortion in current waveforms, thereby leading to flow of harmonics in the power system [6]. To formalize this observation, harmonics describe the distortion of a voltage or current waveform. Each harmonic component occurs at an integer multiple of the fundamental frequency and by leveraging the Fourier transformation, the steady state waveform can be written as

$$f(t) = \sum_{n=1}^{\infty} I_n \sin\left(\frac{n\pi t}{T}\right)\,, \tag{3.5}$$

where $f(t)$ represents the time domain function, n is the harmonic number, and I_n is the magnitude of the nth harmonic, and T is the cycle duration in seconds.

3 This is due to the fact that transformer aging depends highly on the status of the internal insulation material.

Moreover, the aggregate harmonic distortions TDH_i for current and TDH_V for voltage, which are commonly used in the literature, are given by

$$\mathrm{THD}_I \triangleq \frac{1}{I_1}\sqrt{\sum_{n=2}^{N} I_n^2}\,, \tag{3.6}$$

$$\mathrm{THD}_V \triangleq \frac{1}{V_1}\sqrt{\sum_{n=2}^{N} V_n^2}\,, \tag{3.7}$$

where I_1 and V_1 correspond to the root mean square (RMS) values of the main frequency component, I_n and V_n are the RMS-value of the $n\mathrm{th}$ harmonic component, and N is the highest index of the observed harmonic. In the case of PEV charging, the harmonics are related to the charger topology and the underlying power electronics. In most parts of the world there exist regulations for public power supplies that determine the allowable harmonic distortion. For instance, the European Standard, EN 50160, sets the limit of 8% for low voltage supplies and in Malaysia the THD_V limits for 11-kV and 33-kV systems are set as 3% and 4%, respectively. Therefore, such constraints on the harmonic distortions pose certain limitations on the design of the charging stations [7]. There exist real-world empirical studies to further assess this issue. The study in [8] analyzes the harmonics introduced by PEV charging and shows that the total harmonic distortions are $\mathrm{THD}_I = 11.6\%$ and $\mathrm{THD}_V = 1.5\%$. Another study presented in [9] for a 19-kWh Li-ion battery with a 3-kW charger shows that THD_i fluctuates in the range of 4.3% to 4.6 %.

The disruptive effects of harmonic distortion include power loss, heating in power system equipment such as cables and transformers, and malfunctioning of the protective relays. In the United States, major parts of the distribution network consist of aging equipment and the system is not designed to address the charging needs of PEVs. However, while there exist pressing needs for modernizing the equipment, the capital-intensive updates might take up to a few decades. For instance, the United States distribution grid is designed to serve 3-5 houses per transformer, where typically, five houses are served by a 37.5-kVA transformer. In such typical settings, if two level 2 chargers connect to the grid concurrently during peak hours, the local transformer quickly becomes overloaded. Moreover, distribution transformers will experience additional winding and core losses due to the harmonics introduced by PEV charging [10].

3.1.1.3 Distribution System Losses

Distribution system losses are defined as the difference between the amount of electricity entering and leaving the network. There are several technical factors, such as hysteresis, eddy currents, dielectric losses, and line resistance and current, that lead to power losses at the distribution network. Controlling the losses is imperative, as such losses directly translate into financial losses, which are imposed, primarily, on the customers. Some recent studies show that uncontrolled PEV penetrations increase the distribution system losses [11]. Therefore, the relationship between the PEV load and the losses should be carefully considered. This relationship is delineated by the Buller and Woodrow formula, which can be described as follows;

$$\text{LSF} = (1 - C)(\text{LF})^2 + C(\text{LF}) , \tag{3.8}$$

where the parameters involved are [11]:

- LSF is the loss factor, defined in the range $[0, 1]$;
- LF is the load factor, defined in the range $[0, 1]$; and
- C is a constant between 0.15 and 0.30.

Recall that the LF is the ratio of the average load to the peak load over a given period. Since the load is proportional to the current, LF can be written as

$$\text{LF} = \frac{I_{\text{average}}}{I_{\text{peak}}} , \tag{3.9}$$

where I_{average} and I_{peak} are the average and the peak currents, respectively. In order to compute the total loss energy, we define $L_{\max}$ as [11],

$$L_{\max} = K \cdot I_{\text{peak}}^2 , \tag{3.10}$$

where K is a positive constant related to the underlying distribution line. Therefore, the total energy losses over time interval T can be calculated according to

$$L_{\text{tot}} = \text{LSF} \times L_{\max} \times T . \tag{3.11}$$

As a result, minimizing the peak current reduces the loss factor and the total energy loss. Hence, based on this formulation, the charging strategies should aim to maximize the load factor to minimize energy losses.

3.1.1.4 System Unbalances

System unbalances usually happen when the loads are not uniformly distributed across the feeder. Considering the sizes of nonlinear PEV batteries, the penetration of PEVs can unbalance a three-phase system, meaning that voltage and current in one phase differ from those of the other phases. In such circumstances, the phase difference deviates from the standard 120°. System unbalances impose additional losses and heating in the power system components. In three-phase systems, the unbalance can appear as a sum of positive, negative, and zero voltage sequences. The voltage unbalance has been described by organizations such as the Institute of Electrical and Electronics Engineers (IEEE), National Electrical Manufacturers Association (NEMA), International Electrotechnical Commission (IEC), and International Council on Large Electric Systems (CIGRE) [12].

IEEE Definition

According to the IEEE Recommended Practice for Monitoring Electric Power Quality, phase voltage unbalance rate (PVUR) in percentage is calculated according to

$$\mathrm{PVUR} = \left| \frac{V_{\mathrm{phase}}^{\mathrm{max}}}{V_{phase}^{\mathrm{avg}}} \right| \times 100 \,, \tag{3.12}$$

where $V_{\mathrm{phase}}^{\mathrm{max}}$ is the maximum voltage deviation from the mean phase voltage magnitude, and $V_{\mathrm{phase}}^{\mathrm{avg}}$ is the mean phase voltage magnitude [13].

NEMA Definition

Similar to the IEEE definition, the NEMA voltage definition uses the measurement of voltage magnitude to calculate the line voltage unbalance rate (LVUR), which is given by

$$\mathrm{LVUR} = \left| \frac{V_{\mathrm{line}}^{\mathrm{max}}}{V_{line}^{\mathrm{avg}}} \right| \times 100 \,, \tag{3.13}$$

where $V_{\mathrm{line}}^{\mathrm{max}}$ is the maximum voltage deviation from the mean line voltage magnitude and $V_{\mathrm{line}}^{\mathrm{avg}}$ is the mean line voltage magnitude. Notice that the only difference from the IEEE definition is that NEMA uses line voltages whereas IEEE uses phase voltages.

IEC Definition

The IEC definition is typically used in the European standards and defines the voltage unbalance factor (VUF) according to

$$\mathrm{VUF} = \frac{V_2}{V_1} \times 100\,, \tag{3.14}$$

where V_1 and V_2 are defined as follows with phase indices a, b, and c,

$$V_1 = \frac{V_{ab} + aV_{bc} + a^2 V_{ca}}{3}\,, \tag{3.15}$$

and

$$V_2 = \frac{V_{ab} + a^2 V_{bc} + a^2 V_{ac}}{3}\,. \tag{3.16}$$

CIGRA Definition

The CIGRA provides a simple algebraic definition for VUF given by

$$\mathrm{VUF} = \sqrt{\frac{1 - \sqrt{3 - 6\alpha}}{1 + \sqrt{3 - 6\alpha}}} \times 100\,, \tag{3.17}$$

where α is defined as

$$\alpha = \frac{V_{ab}^4 + V_{bc}^4 + V_{ca}^4}{(V_{ab}^2 + V_{bc}^2 + V_{ca}^2)^2} \tag{3.18}$$

3.1.1.5 Case Study on Voltage Profiles

We provide a case study on uncontrolled PEV charging to show the impacts on the distribution grid. Details of the study are given in [14]. The vehicles are assumed to be randomly connected at an IEEE 34-node test feeder and the maximum battery capacity is selected as 11 kWh. Moreover, the charger technology was assumed to be uniform across the distribution grid and it has a maximum charging power of 4 kW[4].

4 The maximum charging power for a single phase 230 volt outlet is 4.6 kW; therefore, 4 kW is an appropriate rating.

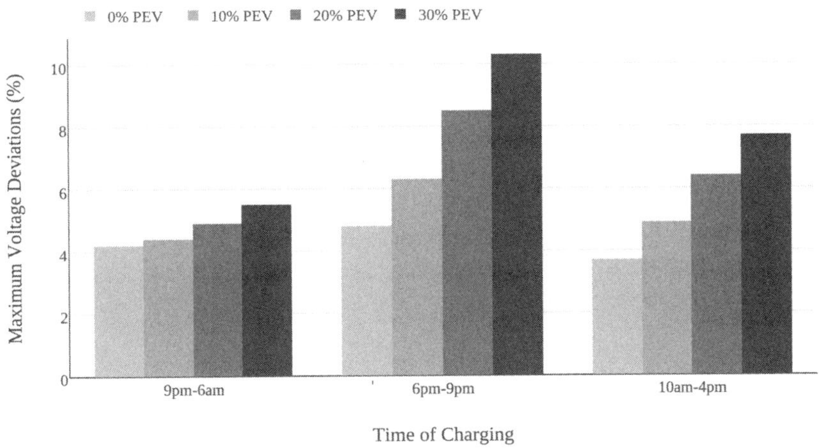

Figure 3.1 Voltage deviations (%) for winter load profiles [14].

The case study considers three different charging periods: (1) 9 pm-6 am; (2) 6 pm-9 pm; and (3) 10 am-4 pm. This case study is conducted in a Belgian distribution grid by incorporating load profiles from the same region. Therefore, the results will differ in other countries. The simulation methodology is as follows. Based on three different penetration scenarios, 0%, 10%, 20%, and 30%, PEVs are randomly located at each node and load flow analysis is conducted with the backward-forward sweep methodology. The simulation is repeated 1000 times to achieve steady-state averages for voltage deviations and power losses.

The simulation results presented in Figures 3.1 and 3.2 show the effects of uncontrolled charging on the voltage deviations and power losses. It is noteworthy that if the charger ratings were higher (e.g., level 2 chargers with 6.6 kW or higher) the impacts would have been more severe. As the results suggest, the losses are high during peak hours and this is reflected back to customers via increasing electricity tariffs. [5]

5 Detailed analyses of utility tariffs are presented in the next chapters.

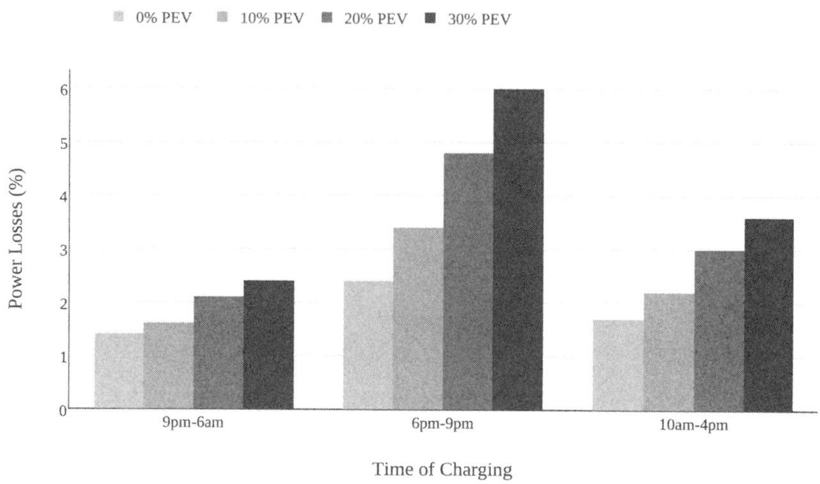

Figure 3.2 Ratio of power losses to maximum power for winter load profiles [14].

3.1.2 Impacts on Generation Portfolio

The impacts of PEV demands on the power generation, such as demand peaks, reduction in reserve margins, and increased electricity prices, depend on the regional penetration rates, the charging technology deployed, and the time of charging. The studies in [15] suggest that PEV demands can increase generation and electricity prices if they PEVs are served during peak hours. Obviously, the real impacts also depend on the penetration rates. For instance, if all PEVs, which are assumed to be 25% of the vehicle fleet, are charged at the same time by using level 2 chargers, the demand increases by 160 GW by 2030. The same study also analyzes the regional demand increases in, for instance, the Electric Reliability Council of Texas (ERCOT), Southeast Electric Reliability Council (SERC), and Southwest Power Pool (SPP), and shows that demand increase can be as high as 4.7%. Overall, uncontrolled PEV charging decreases the utilization of low-cost resources such as nuclear and hydro, it is likely to increase the peak to average load ratio, and consequently, increases the cost of energy generation. Therefore, it is necessary to have intelligent mechanisms to schedule PEVs during off-peak hours to prevent peak rises. Another benefit of charging PEVs during off-peak hours is that PEV demands increase the utilization of low-cost generators, which in return lowers generation cost.

3.1.3 Transmission Network

The transmission network is the most advanced and upgraded part of the power grid. However, transmission lines often experience congestion during high demand periods. For instance, statistics from the United States Department of Energy indicate that at least 17% of the Western Interconnection network is congested for at least 10% of the time. Therefore, the increase in the demand may lead to even more congestions in the transmission network. Hence, avoiding bottlenecks due to transmission lines necessitates mechanisms for controlling and scheduling PEV demands.

3.2 Network of Charging Stations

The adoption rate of PEVs strongly hinges on the availability of charging networks in order to cope with an array of challenges, including the range anxiety of drivers. Moreover, the convenience of the service through timely charging with controlled waiting and charging delays, or lack of these, can influence the adoption rate. Addressing such issues, at its core, requires a network-level design for the charging infrastructure in order to balance the demand and response, and to optimally schedule the charging resources while satisfying constraints on consumers' preferences. A charging network consists of three main charging methods, which are, as described in the previous chapter, level 1 (e.g., garage) charging at the street level, level 2 charging at the workplace, hospitals, or shopping malls, and DC fast charging stations, which are the counterparts of the current gas stations for PEVs. Garage charging is mainly preferable for customers who have access to private parking. However, PEV owners living in large metropolitan areas may not have convenient or viable access to this option. As a remedy, during the weekdays, PEVs can use level 2 charging stations available in public places. One advantage of the level 2 charging option is that PEV owners have the option of charging the car while being in a stationary or parked mode for moderate to long periods of time, during which they do not need the vehicle. Such charging stations are also likely to be equipped with renewable resources and will be morc environmentally conscious. The third charging option includes the fast-charging stations, which can charge a typical PEV in 20 to 30 minutes. This type of station has a critical role in expanding availability

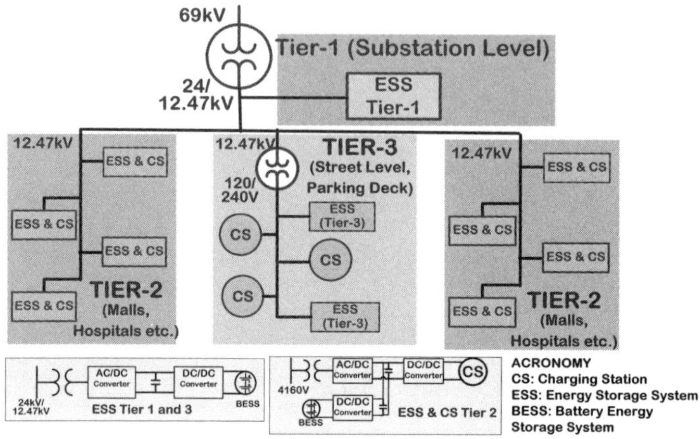

Figure 3.3 Illustration of a charging station network equipped with energy storage units.

of the charging stations to the same extent that gas-powered vehicles have access to gas stations. Moreover, DC fast stations enable drivers to plan long-distance trips without experiencing any discomfort due to long charging service.

A representative illustration of a three-tier charging network is presented in Figure 3.3. In the model, a large-scale MWhr-rated storage unit is deployed at a typical 12-kV point of common coupling (PCC). At the second tier, level 2 chargers are incorporated into the grid, which can serve customers at large complexes such as university campuses, airports, and shopping malls. A typical size of the storage unit is around 100 to 250 kWhr. At the third tier, slow chargers are incorporated to serve the representative system showing that the operation of the charging networks is a complex problem. In addition to a multi-tier structure, the following issues should also be accounted for in the designs:

- Dynamically matching the constantly varying *spatiotemporal* demands with the limited grid resources.
- Efficient operation of individual charging stations, possibly by enhancing their operations via equipping them with energy storage units and renewable generators.
- Optimal control, scheduling, and optimization frameworks to form a closed-loop system in which the available resources can also shape the arrival rates of the customer demands.

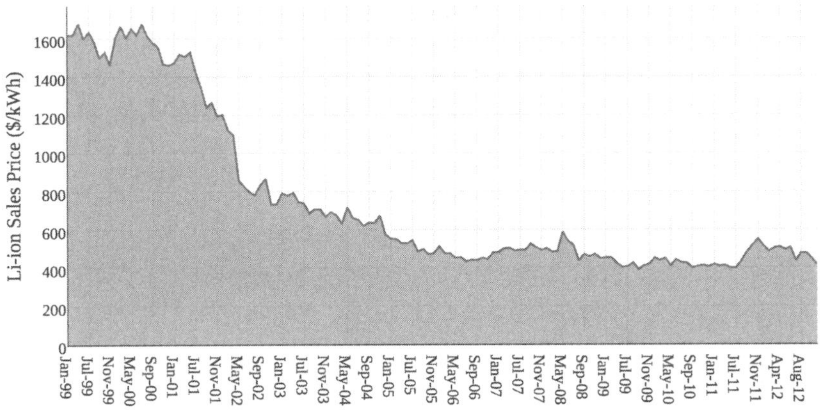

Figure 3.4 Historical cost of Li-ion batteries from 1999 to 2012. Prices are shown in 2012 dollars.

- Capacity planning and load forecasting to facilitate look-ahead planning.
- Load balancing and customer routing framework among the charging stations to alleviate the impacts of nonuniform customer demand.

3.3 High Acquisition Cost

While PEVs are expected to have lower fuel and operating expenses, their acquisition cost is relatively higher due to the cost of batteries and the associated technologies. A comprehensive analysis and comparison on the return on investment (ROI) for different vehicle classes depend on several key factors including driving patterns, electricity generation cost, and gas price.

Even though there is no consensus on the relative costs of the technologies, there is a widely agreed-upon understanding that the acquisition cost of PEVs, on average, is about $5000-$8000 higher than that of the gas-powered vehicles. This difference is primarily due to the cost of batteries. Even though battery cost is decreasing, as shown in Figure 3.4, the technology still requires significant improvements to increase the all-electric-range and reduce the initial acquisition

Table 3.1
PEV Incentives around the World

Country	Incentive
Austria	Exemption from fuel and monthly vehicle tax.
Belgium	Exemption from registration tax. Deductions from corporate income tax.
China	Rebates about 2200-7200 EUR.
Germany	Exemption from annual circulation tax.
Denmark	Exemption from registration tax.
France	Offered rebates from 2000 to 6300 EUR depending on emission levels.
Italy	Exemption from the annual circulation tax.
Japan	Rebates up to 6300 EUR.
Netherland	PEVs are exempt from the registration tax.
Norway	Exemption from VAT and purchase tax could add up to 50% of the cost.
Sweden	Tax exemptions for the first five years
United Kingdom	Rebates up to 5000 GDP. Exemption from circulation tax.
United States	Rebates up to 7500 USD. Additional incentives are provided by states.

cost. In the meantime, some governments around the globe are providing incentives to promote PEV deployment. The incentive programs, broadly, can be divided into three groups:

- Direct incentives, which are in the form of a one-time bonus (or discount) upon the purchase.
- Fiscal incentives, which are in the form of tax exemptions.
- Miscellaneous incentives such as emission test exemption, free parking, license fee reduction, high-occupancy vehicle (HOV) lane access, free charging, and rebates for in-home chargers.

For instance, in the United States, through the American Recovery and Reinvestment Act of 2009, the Federal Government offers $2500 – $7500 tax credits depending on battery size. State governments are also providing incentives to promote PEV penetration. An overview of incentives is provided in Table 3.1. The work in [16] focuses on incentives in the European Union and in Nordic countries. More specifically, the incentives are grouped into three categories:

- *Leader* countries are Denmark, Norway, and Belgium. The total incentives range from 10,000 to 28,000 EUR.
- *Follower* countries are Netherlands, Spain, the United Kingdom, France, Switzerland, and Austria. The total incentives range from 4000 to 9000 EUR.
- *Laggard* countries are Ireland, Greece, Italy, Germany, Sweden, Poland, and Finland. The total incentives are around 3000 EUR.

3.4 Chapter Summary

In this chapter, we have presented the roadblocks that are negatively affecting PEV adoption. The first roadblock is the capability of the power grid to handle extra demand introduced by PEVs. We have discussed impacts of PEV charging especially on the distribution grid, as the negative impacts will be mostly noticed at the distribution level. The uncontrolled charging of PEVs can cause thermal loading, harmonic distortion, and increased distribution system losses. Furthermore, we have discussed the impacts on the generation and the transmission networks. The other two roadblocks are the network of charging stations and high acquisition cost. This chapter has also provided an introduction to the network of charging stations, which will be discussed more in detail in the upcoming chapters. To tackle the high acquisition cost we have presented an overview of incentives for different countries.

References

[1] National Research Council on the National Academies, *Overcoming Barriers to Deployment of Plug-in Electric Vehicles*, Washington, DC: The National Academies Press, 2015.

[2] Kahneman, Daniel. *Thinking, Fast and Slow*, New York: Farrar, Straus and Giroux, 2011.

[3] Adner, Ron, "Match your innovation strategy to your innovation ecosystem", *Harvard Business Review*,Vol. 84, No. 4, 2006, pp. 98-107.

[4] *IEEE Guide for Loading Mineral-Oil-Immersed Transformers and Step-Voltage Regulators - Redline*, in IEEE Std C57.91-2011 (Revision of IEEE Std C57.91-1995) - Redline, Mar., 2012, pp.1-172.

[5] Hilshey, A.D. et al., "Estimating the impact of electric vehicle smart charging on distribution transformer aging," *IEEE Transactions on Smart Grid*, Vol. 4, No. 2, 2014, pp. 905-913.

[6] IEEE Power and Energy Society, *IEEE Recommended Practice and Requirements for Harmonic Control in Electric Power Systems*, IEEE-SA Standards Board, Mar. 2014.

[7] Kutt, L. et al., "A review of the harmonic and unbalance effects in electrical distribution networks due to EV charging," *Proc. IEEE International Conference on Environment and Electrical Engineering*, Wroclaw, Poland, May 2013, pp. 556-561.

[8] Wanik, MohdZamriChe, et al. "Harmonic measurement and analysis during electric vehicle charging," *Engineering*, Vol. 5, No. 1, 2013, pp. 215-220.

[9] Wenge, C. et al., "Power quality measurements of electric vehicles in the low voltage power grid," *Proc. International Conference on Electrical Power Quality and Utilisation*, Oct. 2011, pp.1-5.

[10] Moses, P.S., Masoum, M.A.S., and Smedley, K.M., "Harmonic losses and stresses of nonlinear three-phase distribution transformers serving plug-in electric vehicle charging stations," *Proc. IEEE Innovative Smart Grid Technologies*, Washington, DC, Jan., 2011, pp.1-6.

[11] Sortomme, E. et al., "Coordinated charging of plug-in hybrid electric vehicles to minimize distribution system losses," *IEEE Transactions on Smart Grid*, Vol. 2, No. 1, 2011, pp.198-205.

[12] Singh, Asheesh K., G. K. Singh, and R. Mitra. "Some observations on definitions of voltage unbalance," *Proc. 39th IEEE North American Power Symposium*, Las Cruces, New Mexico, Oct., 2007, pp. 473-479.

[13] IEEE Recommended Practice for Monitoring Electric Power Quality in *IEEE Std 1159-2009 (Revision of IEEE Std 1159-1995)* , June 26, 2009.

[14] Clement-Nyns, Kristien, Edwin Haesen, and Johan Driesen. "The impact of charging plug-in hybrid electric vehicles on a residential distribution grid," *IEEE Transactions on Power Systems*, Vol. 25, No. 1, 2010, pp. 371-380.

[15] Hadley, Stanton W., and Alexandra A. Tsvetkova. "Potential impacts of plug-in hybrid electric vehicles on regional power generation," *The Electricity Journal*, Vol. 22, No. 10, 2009, pp. 56-68.

[16] Wu, Qiuwei. *Grid Integration of Electric Vehicles in Open Electricity Markets*, Chester, UK: John Wiley & Sons, 2013.

Chapter 4

PEV Cost Models

In Chapter 3, we provided an overview of adverse impacts of uncontrolled PEV charging on power grid operations and efficiency. Controlled integration of PEVs addresses those adverse impacts, but also introduces its own challenges, which will be continually addressed in this book. In order to formalize and address these challenges, the primary goal of which is minimizing the service cost while controlling the impacts on the grid, we introduce two more important parameters: *battery degradation* and *electricity cost*. These two parameters are instrumental to delineating PEV charging control and optimization frameworks. Battery degradation cost depends on an array of factors, the predominant ones being charging power, depth-of-discharge, and temperature. The electricity cost, on the other hand, is often controlled by the utility companies based on various pricing schemes, which we review in this chapter.

4.1 Battery Degradation Cost

The first three chapters provided an overview of electric vehicle batteries, and concluded that the batteries are predominantly the most costly component of PEVs. This section aims to provide a more detailed analysis of the aging mechanisms of Li-ion cells and review the mathematical models of battery degradation costs.

Due to their chemical properties, Li-ion batteries are complex systems. Hence, the task of developing aging models is particularly complicated as the battery degradation originates from a variety of intertwined factors that affect the physical properties of the anode, cathode, and electrolyte. Therefore, we will investigate the general aspects of battery aging, the root causes and leads, and the common mathematical models that are widely used in the literature. Overall, battery degradation can occur during energy storage due to, for example, self-discharge and impedance rise, as well as during charging/discharging periods due to, for example, metal plating and mechanical degradation, which are governed by the thermodynamical stability of the battery. Battery degradation during storage affects calendar life, while deterioration during the battery use lowers cycle life [1].

Battery degradation can be viewed in the form of its capacity fading or its power fading. Capacity fade refers to the reduction in the ability to store energy and leads to a decrease in the all-electric range. On the other hand, power fade degrades the output power of the battery and affects vehicle acceleration and top speed. The inherent reason for the power fade is that battery impedance degrades (increases) over time, and subsequently, it limits the current flow.

4.1.1 PEV Battery Modeling

Developing mathematical models to understand the factors that affect battery degradation under realistic driving scenarios has paramount importance. This is especially imperative in the existing technologies as the current battery packs are significantly oversized for the purpose of meeting the energy and power requirements [2]. In the next sections, we review three Li-ion battery degradation cost models that are widely adopted in the recent literature.

4.1.2 Modeling Approach I: NREL Model

The first model developed for battery degradation estimates both capacity fade and power fade of a Li-ion battery as functions of the following three main drivers: (1) temperature, (2) state-of-charge profile, and (3) daily depth of charge. The model is developed primarily based on the data from the National Renewable Energy Laboratory in the United States [3, 4]. Let C represent the cost of purchasing a new battery and define L as the total battery lifetime given by the manufacturer. Additionally, during each cycle, the battery lifetime shrinks by L^{-} units. Therefore, the battery degradation cost C_{deg} can be formalized as

$$C_{\text{deg}} = C \cdot \frac{L^-}{L} \ , \tag{4.1}$$

where $\frac{L^-}{L}$ captures the ratio by which the lifetime shrinks during each cycle. It is noteworthy that degradation calculations are based on the notion of the end of life (EOL), denoted by B_{EOL}, which as discussed in Chapter 2, is generally set as 80% of the initial battery capacity B_0, i.e., $B_{\text{EOL}} = 0.8 \cdot B_0$. The EOL power requirements are typically overprovisioned with respect to energy requirements as power requirements at low state of charge (SoC) levels (e.g., 20%) requires using costly materials. Therefore, EOL power level P is chosen as $P = 0.7 \cdot P_0$, where P_0 is the initial maximum power of the battery pack.

4.1.2.1 Power Fade and Internal Resistance

We start our analysis by developing a model for the time-varying power fading process, denoted by $P_{\text{fade}}(t)$, which occurs due to temporal growth of resistance, as shown by NREL studies [2]. The model is based on the battery test manual published by the Idaho National Laboratory [5], and the calculations are made based on the internal resistance values. Let V_{min} and V_{max} represent the minimum and maximum operating battery voltages specified by the manufacturer (also explained in Chapter 2). Maximum time-varying charge and discharge power values are typically distinct, and they are represented by $P^{\text{ch}}_{\text{max}}(t)$ and $P^{\text{disch}}_{\text{max}}(t)$, respectively. The open circuit voltage is denoted by V_{OC}, which varies by SoC. Also, the internal resistance is denoted by R. Then, from circuit theory, $P^{\text{ch}}_{\text{max}}(t)$ and $P^{\text{disch}}_{\text{max}}(t)$ can be calculated according to

$$P^{\text{ch}}_{\text{max}}(t) = V_{\text{max}} \cdot \frac{V_{\text{max}} - V^{\text{ch}}_{\text{OC}}}{R_{\text{ch}}(t)} \ , \tag{4.2}$$

and

$$P^{\text{disch}}_{\text{max}}(t) = V_{\text{min}} \cdot \frac{V^{\text{disch}}_{\text{OC}} - V_{\text{min}}}{R_{\text{disch}}(t)} \ . \tag{4.3}$$

Therefore, the time-varying power fading process can be approximated by

$$P^{\text{ch}}_{\text{fade}}(t) \approx \frac{P^{\text{ch}}_{\text{max}}(t)}{P_0} \approx \frac{R^{\text{ch}}_0}{R(t)} \ , \tag{4.4}$$

$$\text{and} \quad P^{\text{disch}}_{\text{fade}}(t) \approx \frac{P^{\text{disch}}_{\text{max}}(t)}{P_0} \approx \frac{R^{\text{disch}}_0}{R(t)} \ , \tag{4.5}$$

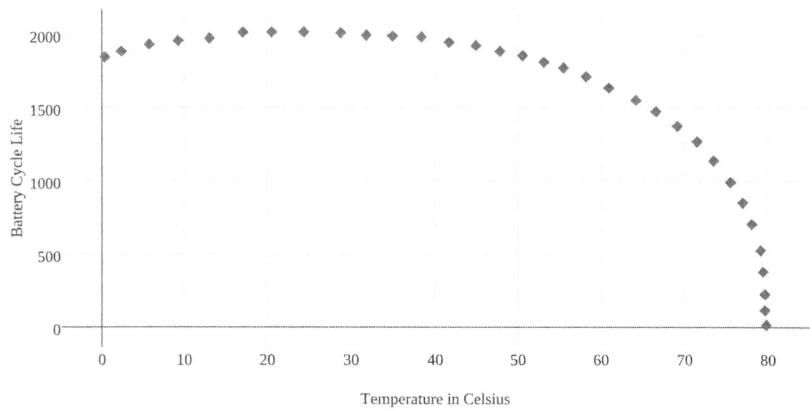

Figure 4.1 Laboratory test for cycle life versus temperature [6].

where P_0 and R_0 are the initial power and the internal resistance of the battery. Equation (4.4) shows that as internal resistance grows, power fade increases and battery performance degrades.

In the next three sections, we analyze how power fade and capacity fade are affected by the operating temperature, open circuit voltage, and depth of discharge (DoD), which are all affected by the growth of the internal resistance of the battery.

4.1.2.2 Battery Degradation Due to Temperature

The Arrhenius relationship, given by

$$r = A \cdot \exp\left(-\frac{E_a}{kT}\right),$$

has long been serving as the primary model to understand the temperature-induced degradation effects in batteries. In the Arrhenius relationship, r accounts for the rate of reaction, E_a is the activation reaction energy, k is the Boltzmann constant, T is the operation temperature, which indicates that kT is the average kinetic energy, and finally A is a constant determined by battery tests [2, 6]. Figure 4.1 shows the relationship between the battery cycle life and operating temperature. The lifetime of the battery as well as power fade and capacity fade are inversely proportional to the rate of the reaction r. Therefore, the lifetime loss at temperature T for power fade is

$$L_p(T) \triangleq a_p \exp(b_p/T) , \tag{4.6}$$

and for capacity fade is

$$L_q(T) \triangleq a_q \exp(b_q/T) . \tag{4.7}$$

Determining and calibrating parameters $\{a_p, a_q, b_p, b_q\}$ are often carried out in accordance with the available data through data fitting methods [2, 7]. Calculating battery degradation over a given time period requires incorporating the exact operating temperatures. It is noteworthy that the battery life L, which is leveraged for defining battery degradation cost in (4.1), is determined based on the charging scenario provided by the manufacturers. Such scenarios are usually described based on the minimum charging power denoted by $P^{\rm ch}_{\rm min}$. Next, we review three settings in which the battery degradation can occur.

- The battery can be charged at a power higher than $P_{\rm min}$, hence the operating temperature is $T_1 = T_{\rm amb}+R(t)\cdot P(t)$, where $T_{\rm amb}$ is the ambient temperature, $R(t)$ is the internal resistance, and $P(t)$ is the charging power.
- The battery can be plugged in but not being charged, in which case the relationship between the plug-in time $t_{\rm plug}$ and charging duration $t_{\rm ch}$ is $t_{\rm ch} \leq t_{\rm plug}$. In this setting the operating temperature is $T_2 = T_{\rm amb}$.
- The battery can be charged at the minimum allowable power, which also serves as a baseline case and the operating temperature is $T_3 = T_{\rm amb} + R(t) \cdot P_{\rm min}$.

Hence, the additional power fade due to temperature during a year (i.e., $h_{\rm tot} \triangleq 8760$ hours) is

$$C_{{\rm deg},T,p} = C \cdot \left\{ \int\limits_{t_{\rm ch}} \frac{1}{h_{\rm tot}} L_p(T_1) dt + \frac{t_{\rm plug} - t_{\rm ch}}{h_{\rm tot} L_p(T_2)} - \frac{t_{\rm plug}}{h_{\rm tot} L_p(T_3)} \right\} . \tag{4.8}$$

Similarly, the capacity fade due to temperature becomes

$$C_{{\rm deg},T,q} = C \cdot \left\{ \int\limits_{t_{\rm ch}} \frac{1}{h_{\rm tot}} L_q(T_1) dt + \frac{t_{\rm plug} - t_{\rm ch}}{h_{\rm tot} L_q(T_2)} - \frac{t_{\rm plug}}{h_{\rm tot} L_q(T_3)} \right\} . \tag{4.9}$$

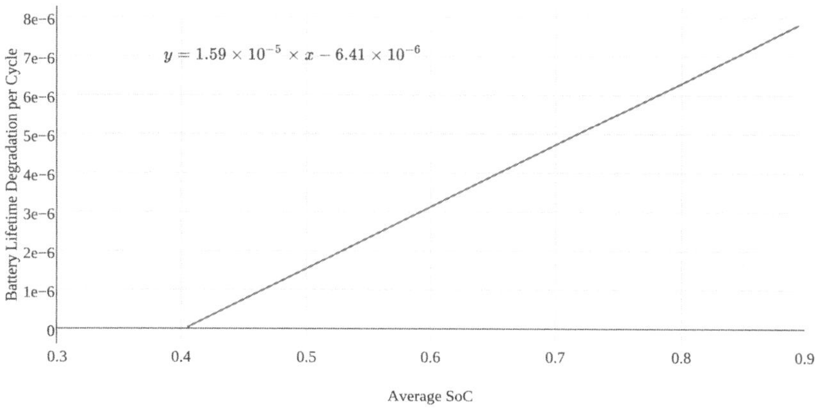

Figure 4.2 SoC related battery degradation based on NREL model data [3].

4.1.2.3 Battery Degradation Due to State-of-Charge

State-of-charge (SoC) related battery degradations, denoted by $C_{\text{deg,SoC},q}$, affect only the battery capacity. As the battery SoC approaches its lower and upper extreme values, the internal resistance increases, hence the battery capacity fades over time. However, as explained in Chapter 2, accurately estimating the SoC related degradations is intractable, and it is often approximated by leveraging the historical experimental data. Based on the data from NREL, degradation can be specified according to

$$C_{\text{deg,SoC},q} = C \cdot \frac{a \cdot \text{SoC}_{\text{avg}} - d}{QF_{\text{max}} \cdot y \cdot h_{\text{tot}}}, \tag{4.10}$$

where a and d are the linear fitting parameters, QF_{max} is the capacity fade at EOL (e.g., 20%) and y is time in years. Figure 4.2 demonstrates an example in which the linear relationship between $C_{\text{deg,SoC},q}$ and SoC_{ave} is depicted.

4.1.2.4 Battery Degradation Due to DoD

Degradations due to depth of discharge constitutes the third component of the NREL model, and it is represented by $C_{\text{deg,DoD},q}$. Similar to the SoC case, the major degradation due to DoD occurs in the form of capacity fade. DoD refers to the capacity losses due to SoC swings; that is, the range in which the battery is cycled. For instance, the amount of battery degradation induced by charging a battery between 60%-100% SoC is different from charging the same battery between 50%-90% SoC. Hence, the effects of DoD on battery life cycle can be stated as

$$N(\text{DoD}) \triangleq \left(\frac{\text{DoD}}{145.71}\right)^{-1/0.6844} . \tag{4.11}$$

This relationship is characterized based on the NREL's data provided in [8], where $N(\text{DoD})$ measures battery's lifetime in terms of cycles. In order to compute the degradation level of battery capacity we need to calculate the decrease in the energy output of the battery when it operates under the optimal SoC swing. Let E_{TL} denote the lifetime energy throughput of the battery. [1] Then E_{TL} under an average DoD_{mean} can be written as $E_{\text{TL}} = N(\text{DoD}_{\text{mean}}) \cdot \text{DoD}_{\text{mean}} \cdot C$. Similarly, the amount of energy output, when operated by the manufacturer DoD_{base}, can be stated as $E_{\text{base}} = N(\text{DoD}_{\text{base}}) \cdot \text{DoD}_{\text{base}} \cdot C$. Let E_{used} denote the total change in the energy throughput after a specific cycle DoD_i. The energy throughput with respect to mean DoD_{mean} slightly changes to $\text{DoD}_{\text{mean,i}}$ and the total energy used in this specific cycle equals to $E_{\text{used}} = N(\text{DoD}_i) \cdot \text{DoD}_{\text{mean,i}} \times C + \text{DoD}_i \cdot C$. Finally, the battery degradation can be written as

$$C_{\text{DoD},q} = C \cdot \left(\frac{E_{\text{used}} - E_{\text{base}}}{E_{\text{TL}}}\right) . \tag{4.12}$$

The aggregate degradation can be computed by incorporating the contributions of degradations due to temperature, SoC, and DoD; that is

$$C_{\text{deg}} = C_{\text{deg},T,q} + C_{\text{deg,SoC},q} + C_{\text{DoD},q} . \tag{4.13}$$

4.1.3 Modeling Approach II

The model discussed in Section 4.1.2.4 provided analytical tools to assess battery's lifetime. While being effective and important to understand the underlying complex

1 Lifetime energy throughput is the total amount of energy (in W-hr) that a battery can provide during its lifetime.

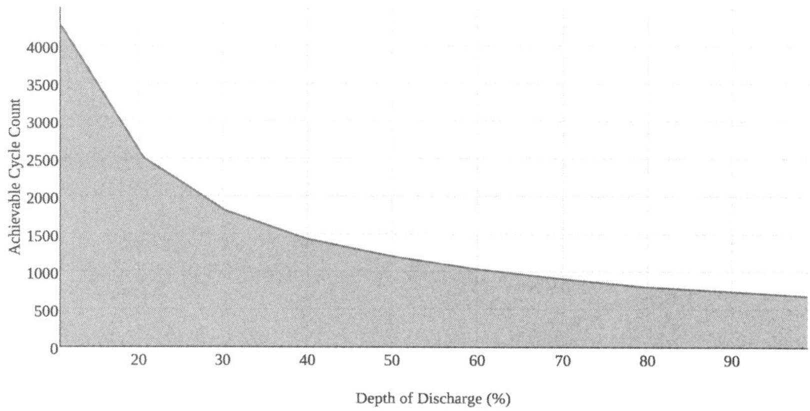

Figure 4.3 Typical Li-ion DoD-ACC model [9].

dynamics, this model depends heavily on the experimental data provided by the NREL and lacks sufficient versatility to be applied to different battery technologies. Additionally, the cost of performing similar experiments for different battery technologies is prohibitive, which necessitates alternative modeling approaches.

The second model presented in this chapter uses the DoD achievable cycle count (ACC) data obtained from battery manufacturers [9]. The advantage of this approach is that the DoD-ACC data is furnished by the manufacturers and does not depend on further external experiments or information. The underlying idea in this model is to develop a degradation density function (DDF) that calculates the total battery degradation cost based on the given DoD-ACC information. The drawback of this model is that it assumes that the battery is charged at a low charging power (e.g., level 1 home charging), and consequently, it ignores the effects of ambient temperature.

In Figure 4.3, we present a typical DoC-ACC graph for a lithium-ion battery for PEVs. The first step in our analysis is to translate this graph into average degradation cost (ADC). Clearly, ADC depends on the DoD, cost of battery C, battery size B_0, and the efficiency at DoD, denoted by ϵ. Specifically, $\mathrm{ADC}(\mathrm{DoD})$ measures battery degradation cost for transferring 1 kWh to increase the SoC level by DoD. For instance, $\mathrm{ADC}(30)$ shows the battery degradation cost for transferring

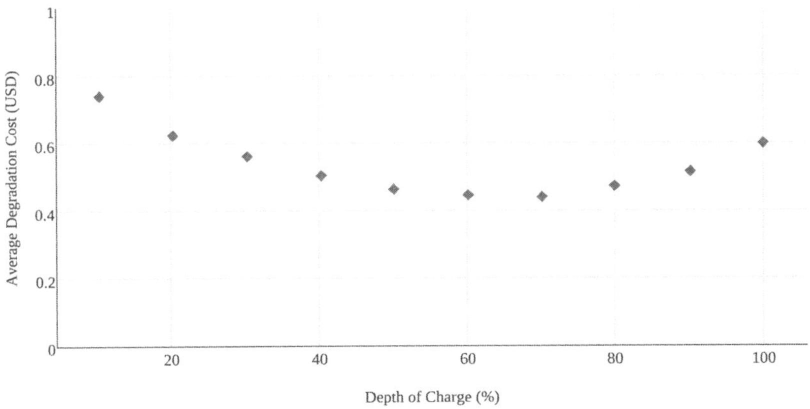

Figure 4.4 A typical Li-ion DoD-SCC model [9].

1 kWh from 70% SoC to 100% SoC. Variation of ADC with respect to DoD is governed by the following relationship:

$$\mathrm{ADC(DoD)} = \frac{C}{2 \cdot \mathrm{ACC(DoD)} \cdot B_0 \cdot \mathrm{DoD} \cdot \epsilon^2} \,. \tag{4.14}$$

In the above equation, ADC is penalized by a factor to account for both the charging and discharging processes in each cycle. Also, charge and discharge efficiencies are assumed to be the same, leading to the quadratic term ϵ^2. Figure 4.4 demonstrates the average degradation cost for a 16-kWh battery that costs 10,000 USD and enjoys 95% efficiency, using the ACC-DoD relationship depicted in Figure 4.3. Recalling the findings in the previous section, the best DoD cycle exhibits at 70% due to the fact that the internal resistance is high at both ends of the SoC.

The drawback of ADC(DoD) is that it always assumes that the cycle starts from 100% SoC, which does not always match the reality. As a remedy, we introduce the degradation density function DDF(SoC) in order to capture the degradation cost for all SoC levels. Based on such a density function, the ADC(DoD) is modified to

$$\text{ADC}(\text{DoD}) = \frac{1}{\text{DoD}} \int_{1-\text{DoD}}^{1} \text{DDF}(\text{SoC}) d(SoC) \,. \tag{4.15}$$

differentiating both sides with respect to DoD yields

$$\text{DDF}(1-\text{DoD}) = -\frac{C}{2 \cdot B_0 \cdot \epsilon^2 \cdot (\text{ACC}(\text{DoD}))^2} \cdot \frac{\partial \text{ACC}(\text{DoD})}{\partial \text{DoD}} \,. \tag{4.16}$$

Given this general form, by leveraging proper curve fitting, an exact ACC-DoD relationship can be obtained. For instance, by selecting the form $\text{ACC}(\text{DoD}) = \frac{x}{\text{DoD}^y}$ for the plot in Figure 4.3, the expression in (4.16) can be written as

$$\text{DDF}(\text{SoC}) = \frac{C}{2 \cdot B_0 \cdot \epsilon^2} \cdot \frac{y \cdot (1-\text{SoC})^{y-1}}{x} \,, \tag{4.17}$$

where parameters x and y are found to be 694 and 0.795, respectively. Plugging these parameters back in (4.16) yields [9]

$$\text{DDF}(\text{SoC}) = 0.322 \cdot \left(\frac{1}{1-\text{SoC}}\right)^{0.205} \,. \tag{4.18}$$

The final step in the analysis is to derive the total battery degradation cost expression for any arbitrary SoC levels, that is;

$$C_{\text{deg}} = B_0 \cdot \int_0^{\mathcal{T}} \text{DDF}(\text{SoC}(t)) \frac{d\text{SoC}(t)}{dt} dt \,, \tag{4.19}$$

where t is the time index, $\mathcal{T}$ is the charging horizon, and $\frac{d\text{SoC}(t)}{dt}$ is the charging pattern (power trajectory). For simplicity, we assume that the PEV is charged with constant power P_c; that is,

$$\frac{d\text{SoC}(t)}{dt} = \frac{1}{B_0} P_c \,. \tag{4.20}$$

By defining $\mathrm{SoC_{arb}} \triangleq \mathrm{SoC}(0)$, the total battery degradation for an arbitrary $\mathrm{SoC_{arb}}$ level is

$$C_{\text{deg}} = \left(-\frac{0.322 \cdot P_c^2}{0.795 \cdot B_0}\right) \cdot \left(\left(1 - \mathrm{SoC_{arb}} - \frac{P_c}{B_0} \cdot \mathcal{T}\right)^{0.795} - (1 - \mathrm{SoC_{arb}})^{0.795}\right) . \tag{4.21}$$

4.1.4 Modeling Approach III

The third modeling approach is a battery-specific approach and aims to characterize the variations of battery degradation with respect to the charging power. Therefore, this approach is more suitable for fast-charging applications. This model is based on the $LiFePO_4$ [2] battery, which is widely used in the current PEV market. [3] In this approach, the goal is to model voltage and current profiles for $LiFePO_4$ and characterize a battery health model. Specifically, the model adopts evolutionary algorithms to estimate a polynomial relationship between battery health on the one hand, and voltage and current values on the other hand, by leveraging the Doyle-Fuller-Newman model. In this approach, the main idea is to simulate and repeat the experiments in multiple rounds in order to get as much information as possible so that the resulting polynomials accurately approximate the underlying physical structures (see [10–12] and the references therein). The polynomial relationship between the battery health, denoted by $H_{\text{cell}}(I, V)$, and the voltage and the current is as follows:

$$H_{\text{cell}}(I, V) = \alpha_1 + \alpha_2 I + \alpha_3 V + \alpha_4 I^2 + \alpha_5 V^2 + \alpha_6 IV + \alpha_7 V^3 . \tag{4.22}$$

The references mentioned above select $\alpha_1 = -0.1148 \cdot 10^{-8}$, $\alpha_2 = 3.9984 \cdot 10^{-8}$, $\alpha_3 = 0.13158 \cdot 10^{-8}$, $\alpha_4 = 554.87 \cdot 10^{-8}$, $\alpha_5 = -4.968 \cdot 10^{-8}$, $\alpha_6 = -1.1166 \cdot 10^{-8}$, and $\alpha_7 = 61.675 \cdot 10^{-8}$. By further assuming that the SoC operates in the ranges of 20%-100%, the charging voltage, V, remains the same as the nominal voltage V_{nom}, i.e., $V = V_{\text{nom}}$. Hence, the charging power P_{cell} can be written as $P_{\text{cell}} = V_{\text{norm}} \cdot I$, and subsequently, (4.22) can be rewritten as

$$H_{\text{cell}}(P_{\text{cell}}) = \beta_1 P_{\text{cell}}^2 + \beta_2 P_{\text{cell}} + \beta_3, \tag{4.23}$$

where, $\beta_1 = \alpha_4 / V_{\text{norm}}^2$, $\beta_2 = \alpha_2 / V_{\text{norm}} + \alpha_6$, and $\beta_3 = \alpha_1 + \alpha_3 V_{\text{norm}} + \alpha_5 V_{\text{norm}}^2 + \alpha_7 V_{\text{norm}}^3$.

2 26,650 cells phospho-olivine structure, continuous 30-C discharge, nominal voltage of 3.3V, operating region of $2 - 3.6$V.

3 Some well-known examples are Chevy Spark, Nissan Leaf, and Renault Clio.

To translate the health degradation cost into a monetary cost, let M represent the number of cells and $g_{\text{cell}}(P_{\text{cell}})$ denote the monetary degradation cost of one cell. Then $g_{\text{cell}}(P_{\text{cell}})$ can be written as

$$g_{\text{cell}}(P_{\text{cell}}) = \frac{C}{M} \cdot V_{\text{norm}} \cdot H_{\text{cell}}(P_{\text{cell}}) \cdot \Delta\mathcal{T}, \tag{4.24}$$

where $\Delta\mathcal{T}$ is the charging duration. This model assumes that all cell units share the same SoC levels. Such an assumption, however, is not always necessarily valid, especially when the batteries age. Nevertheless, even the conclusions from such simplified models are instrumental to providing insights into how the batteries degrade with varying charging rates. The final step in this model is calibrating the model for the entire battery size. Specifically, the aggregate battery degradation at time t is

$$g(P(t)) \;=\; \gamma_1(P(t))^2 + \gamma_2 P(t) + \gamma_3\,, \tag{4.25}$$

where

$$\gamma_1 \;=\; \frac{C \cdot \Delta\mathcal{T} \cdot \alpha_4}{M^2}\,, \tag{4.26}$$

$$\gamma_2 \;=\; V_{\text{norm}} \cdot \Delta\mathcal{T}\frac{C}{M}\left(\frac{\alpha_2}{V_{\text{norm}}} + \alpha_6\right)\,, \tag{4.27}$$

$$\text{and} \quad \gamma_3 \;=\; V_{\text{norm}} \cdot \Delta\mathcal{T} C \beta_3\,. \tag{4.28}$$

4.2 Electricity Cost

Over the last decade, differentiated pricing schemes have become popular. In such schemes, utilities adopt time-varying tariffs, which are computed dynamically for different periods of the day with the purpose of shaping consumers' demands profile. We provide an overview of the details of the pricing schemes employed in the literature, and an illustrative example is presented in Figure 4.5.

4.2.1 All-in-Rate

All-in-rate (AIR) is the most basic pricing mechanism and refers to the static electricity rates that remain unchanged throughout the day. In this pricing mechanism, utilities compute the cost of electricity for different consumption levels, which is used in conjunction with the intervals associated with these consumption levels, in

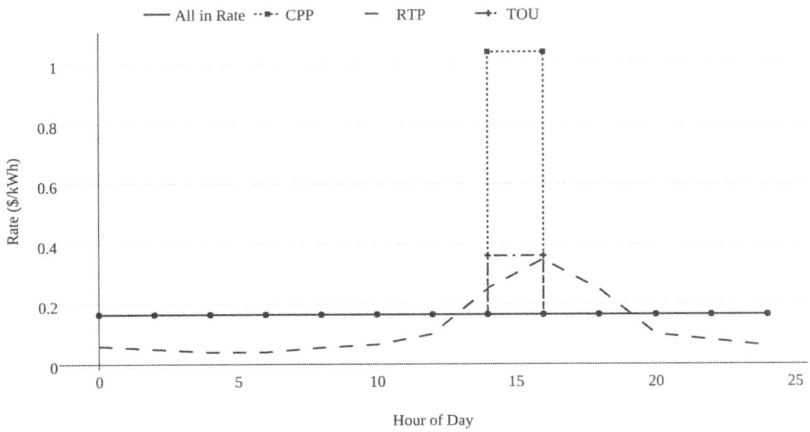

Figure 4.5 Comparison of smart pricing schemes.

order to compute the weighted average cost for each day. Such average costs are then adjusted to account for the electricity delivery cost and other basic surcharges. For instance, in the state of California, there are five pricing tiers, which are defined based on the ratio of the consumption to a baseline power consumption level. Specifically, these five tiers are distinguished by checking to which of the following five intervals the ratio of the actual consumption to the baseline consumption belongs:

$$\{(0-100)\%, (101-130)\%, (131-200)\%, (201-300)\%, (301-\text{above})\%\} ,$$

with a unit cost being $\{0.045, 0.065, 0.151, 0.186, 0.221\}$ USD per kWh, respectively, for these intervals. According to the computations in [13], the average cost per day becomes 0.092 USD per kWh. This cost is adjusted by the average delivery cost (0.072 USD per kWh) and the basic charge (0.020 USD/day), resulting in the all-in summer residential electricity cost of 0.165 USD per kWh. Designing electric rates for commercial and industrial customers is more complicated than for the residential rates. The complexity primarily stems from incorporating the amount and duration of peak consumption.

Based on such models, the average demand charges in the United States for small commercial and industrial customers are 15 and 8 USD kW-month for summer and winter periods, respectively. On the other hand, the demand charges for large commercial and industrial customers are 12 and 8 USD kW-month for summer and winter periods, respectively.

4.2.2 Time-of-Use Rates

In time-of-use (TOU) pricing schemes, days are typically divided into several periods (e.g., peak and midpeak) and for each period a rate is computed. Such periods, and subsequently, the associated prices, depend on the season (e.g., summer versus winter) and the day of the week (e.g., weekends versus weekdays). Clearly, the prices are higher during peak periods and lower during the rest of the time to reflect the cost of electricity generation. TOU rates are held constant throughout every period; therefore, they do not fall into dynamic prices. The very first step in designing the TOU is to set the off-peak electricity price to the utility's marginal cost. Then the peak hour rate is computed to be revenue-neutral to the all-in-rate by considering the average residential load profile. In California, summer season off-peak and peak hour prices are 12.6 cents per kWh and 36.1 cents per kWh, respectively [13]. More rates for other states can be found in [14].

4.2.3 Critical-Peak Pricing

In critical-peak pricing (CPP) schemes during critically loaded periods, customers pay additional peak rates that are higher than other rates. During the rest of the times, however, CPP acts similarly to the TOU pricing scheme. Furthermore, unlike TOU, the peak pricing can change depending on the demand and network conditions. Therefore, this pricing scheme is designed to reflect the true cost of electricity generation. Moreover, end users can save on their consumption costs by following the CPP rates, as the off-peak rates are set lower than the AIR. The CPP is particularly effective during critical days (e.g., hot summer days) during which the stress on the grid reaches its maximum.

The CPP pricing scheme involves three pricing rates: namely critical-peak, off-peak, and peak rate. In order to compute the residential critical-peak price, utilities consider the cost of a typical combustion turbine, and derate this cost by 30% since there is a cost associated with the uncertainties of time of dispatch and the availability of the rate. Next, the derated amount is divided by the number of critically loaded hours, and in the final step, the result is added to the existing all-in rate critical rate. For instance, the typical cost of a combustion engine in California is 75 USD per kW-year and the all-in-rate during summer is 16.5 cents per kWh. Considering 100 hours of critical load during a typical summer season, the corresponding critical-peak rate is $\$\frac{75\times 0.7}{100} + \$0.165 = \$0.69$/kWh. The off-peak rate, however, is roughly set to the marginal electricity generation cost (e.g., \$0.012 /kWh). Then, depending on the customer load profile, the peak rate is computed to maintain revenue neutrality.

4.2.4 Real-Time Pricing

Real-time pricing (RTP) refers to the tariffs that are computed on an hourly basis and this information is delivered to the end users either on hour-ahead or day-ahead bases. Therefore, RTP rates are the closest to the real cost of electricity generation. For instance, in the state of Illinois, two utility companies employ such programs. Customers of Ameren Illinois, who are enrolled in the Power Smart Pricing program, receive hourly prices that are set the night before, and they adjust their usage accordingly. The hourly prices for four different dates are depicted in Figure 4.6. Another Illinois-based utility company called Commonwealth Edison (ComEd), on the other hand, offers an RTP program and broadcasts their prices one hour ahead of dispatch. Rates in RTP for ComEd are calculated according to the formula.$p_d(h) = P_{\text{DA}} + C_{\text{D}} - \text{PI}$, where $p_d(h)$ is the electricity price at hour h on day d, P_{DA} is the day-ahead wholesale market price, C_{D} is the distribution cost (¢5.0/kWh), and PI is the participation incentive (1.4¢/kWh) [15]. Day-ahead prices for four days are depicted in Figure 4.7.

RTP is one of the most powerful dynamic pricing programs. As Figures 4.6 and 4.7 demonstrate, the customers enjoy relatively low prices when compared to other pricing regimes. For instance, the average electricity price for ComEd customers varies in the range of \$0.025 − 0.\$045/kWh, while the U.S. average exceeds \$0.10/ kWh. Therefore, such pricing regimes suggest that PEV charging mechanisms should carefully take pricing information into account.

4.2.5 Inclining Block Rates

Inclining block rate (IBR) is an effective pricing mechanism for avoiding peak demands. Under this pricing policy, the unit price of electricity increases with the usage in blocks of several hundred kWh. Hence, this provides incentives to distribute the usage over time. In South Africa, the prices are $\{0.0469, 0.0502, 0.0621, 0.0717\}$ ¢/kWh for the consumption blocks of $\{(0-50), (52-350), (351-600), (600-\text{above})\}$, respectively. However, this method has not gained much popularity among utility companies.

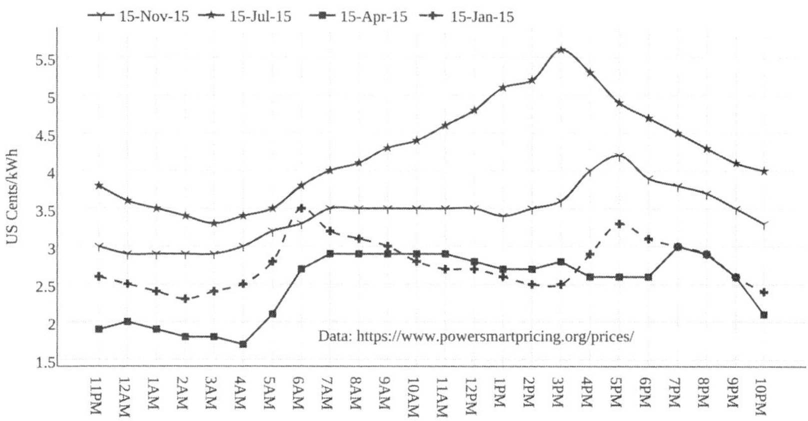

Figure 4.6 Real-time prices, Ameren Illinois.

4.3 Case Study

In this section, we provide a case study to provide a quantitative assessment of how the costs associated with battery degradation and electricity price shape the battery charging profile. This case study relies on the first battery degradation model presented in Section 8.2. For the two cost components we deploy the battery degradation cost, defined in (4.13) and denoted by C_{deg}, and C_{serv}. The total charging service cost paid to the utility company can be computed according to

$$C_{\text{serv}} = \int_{t_{\text{ch}}} C_{\text{elec}}(t)\; P(t)\; dt \;, \tag{4.29}$$

where C_{elec} is the cost of electricity per kWh and, as defined earlier, t_{ch} is the charging duration and $P(t)$ is the charging power. C_{elec} is an exogenous variable given by the utility and can follow one of the pricing tariffs given in the previous section. Subsequently, the total cost becomes

$$C_{\text{tot}} = C_{\text{serv}} + C_{\text{deg}} \;. \tag{4.30}$$

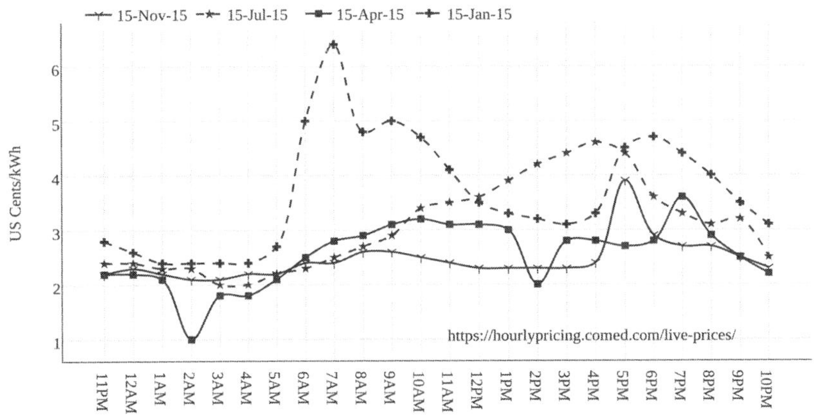

Figure 4.7 Day ahead prices, ComEd Illinois.

This case study represents a potential garage charging scenario, in which the vehicle is stationary and the objective is to find the optimum charging profile that minimizes cost given in (4.30).

Such profile optimization is also subject to constraints pertinent to the available charging power, charting duration, and the state-of-charge. In particular, the charging power is constrained to meet the maximum and minimum charging power constraints according to

$$0 \leq P(t_k) \leq P_{\max} \, . \tag{4.31}$$

It is noteworthy that $P(t)$ can also admit negative values if the vehicle is used in the vehicle-to-grid mode, which can be assessed as an independent case. Furthermore, the charging duration also lies in a pre-specified range (e.g., 24 hours) leading to

$$0 \leq t_{\text{ch}} \leq 24 \, . \tag{4.32}$$

Finally, state-of-charge level is also limited to $0.9t$ in order to protect the battery life, corresponding to which we enforce

$$0.2 \leq \text{SoC} \leq 0.9. \tag{4.33}$$

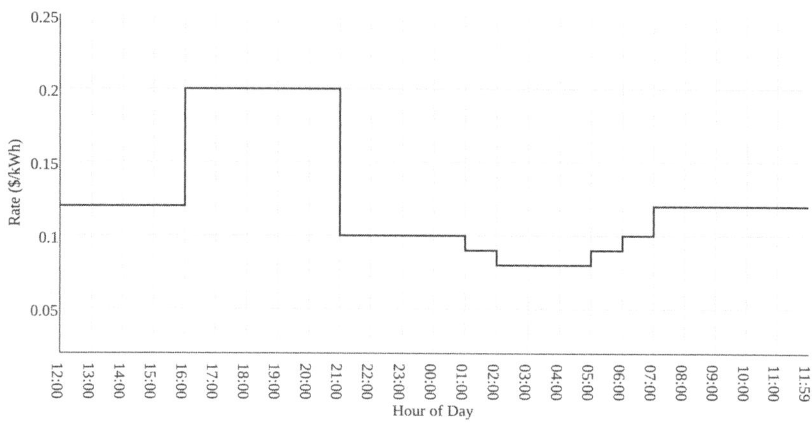

Figure 4.8 Time-of-use pricing.

As a result, the optimization problem can be formalized as

$$\begin{aligned} \text{minimize} \quad & C_{\text{tot}} \\ \text{subject to} \quad & 0 \leq P(t_k) \leq P_{\max} \\ & 0 \leq t_{\text{ch}} \leq 24 \\ & 0.2 \leq \text{SoC} \leq 0.9 \end{aligned} \quad . \tag{4.34}$$

These evaluations are carried out for three PEVs with initial SoC levels of 35%, 30%, and 20%, and each PEV is assumed to have a battery pack of 30 kWh. The maximum charging power is 6.6 kW, the ambient temperature is 25°C, and the thermal resistance is set to 0.002°C/W. Furthermore, we deploy TOU-based pricing, which is presented in Figure 4.8. In this pricing scheme, the electricity unit price drops during nighttime, which is intended to motivate customers to charge their vehicle overnight.

The results are shown in Figure 4.9. When the electricity price is high (e.g., in the afternoon) the vehicle is not charged and the charging power is set to 0. The charging mostly takes place after midnight when off-peak tariffs provide an opportunity for a lowered charging service. This setting also provides a number of

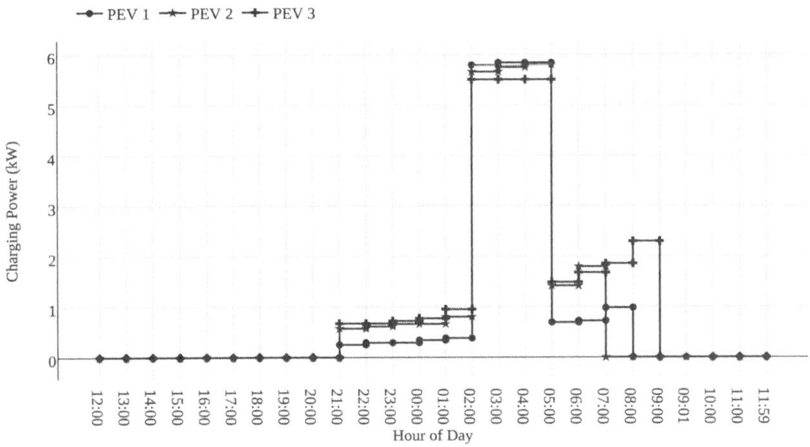

Figure 4.9 Optimum charging power.

insights and raises several research questions, which will be investigated in the next chapters. For instance, the objective function can include more cost components (e.g., a cost can be associated with customers' inconvenience due to delayed service or low quality of service). Moreover, additional constraints can be imposed on the deadline for service completion. For example, in this case study, the service of EV3 is completed at 9 a.m., which may not serve certain needs well. Hence, there exists a trade-off between comfort and cost. Another constraint can be on the usage of shared charging power, in which technological constraints, such as a distribution transformer, may not allow for concurrent charging of multiple vehicles. Furthermore, both the objective function and the constraints can include stochastic variables such as hourly-varying prices or renewable generation, which will introduce additional complexities.

4.4 Chapter Summary

This chapter has concentrated on PEV cost models. Recall that optimal charging of batteries has three major components: power grid operations and limits, Li-ion battery dynamics, and electricity tariffs. Chapter 3 dealt with the limitations of power grids and special attention was given to distribution networks where the negative impacts are expected to be observed the most. The focus of this chapter, on the other hand, had been on PEV battery degradation and the different charging tariffs. As discussed in Chapter 2, PEV batteries represent a significant portion of the cost of PEV ownership, and therefore, the degradation models require special treatment.

This chapter presented three major degradation models that are used in the literature. The first model was developed by NREL and estimated battery capacity and power fade due to ambient temperature, the state of charge profile, and daily depth of charge using the data collected from experimental setups. The second model was an analytical approach and employed depth of charge and achievable cycle count to examine the battery lifetime. The last model was a battery-specific approach for quick-charging applications. In the second part of the chapter, various pricing regimes (e.g., time-of-use, dynamic pricing, and critical peak pricing) were discussed. The case study presented in the last part of the chapter illustrated how pricing and battery degradation determines the charging rate control of PEVs.

References

[1] Vetter, J., et al. "Ageing mechanisms in Li-ion batteries," *Journal of Power Sources*, Vol. 147, No. 1, 2005, pp. 269-281.

[2] Smith, K., Markel, T., Kim, G. H., and Pesaran, A. (2010, October). "Design of electric drive vehicle batteries for long life and low cost." In *Workshop on Accelerated Stress Testing and Reliability.*

[3] Hoke, A., et al., "Accounting for lithium-ion battery degradation in electric vehicle charging optimization," *IEEE Journal of Emerging and Selected Topics in Power Electronics*, Vol. 2, No. 3, 2014, pp. 691-700.

[4] Hoke, A. et al., "Electric vehicle charge optimization including effects of Li-ion battery degradation," *Proc. IEEE Vehicle Power and Propulsion Conference*, Chicago, IL, Sep. 2011, pp. 1-8.

[5] Belt, Jeffrey R. *Battery Test Manual for Plug-in Hybrid Electric Vehicles*, No. INL/EXT-07-12536, Idaho National Laboratory (INL), 2010.

[6] Zhou, C., et al., "Modeling of the cost of EV battery wear due to V2G application in power systems," *IEEE Transactions on Energy Conversion*, Vol. 26, No. 4, 2011, pp. 1041-1050.

[7] Millner, A., "Modeling lithium ion battery degradation in electric vehicles," *Proc. IEEE Innovative Technologies for an Efficient and Reliable Electricity Supply Conference*, Waltham, MA, Sep. 2010, pp. 349-356.

[8] Rosenkranz, C. ,"Deep-cycle batteries for plug-In hybrid application," *EVS-20 International Electric Vehicle Symposium and Exhibition*, Long Beach, CA, Nov. 2003.

[9] Han, S., Han S., and Aki H.. “A practical battery wear model for electric vehicle charging applications,” *Applied Energy*, Vol. 113, 2014, pp. 1100-1108.

[10] Forman, J., Stein, J., and Fathy, H.“Optimization of dynamic battery parameter characterization experiments via differential evolution”, *Proc. IEEE American Control Conference*, Washington, DC, Jun. 2013, pp. 867-874.

[11] Ma, Z., Shichang Z., and Liu, X.“A distributed charging coordination for large-scale plug-in electric vehicles considering battery degradation cost,” *IEEE Transactions on Control Systems Technology*, Vol. 23, No. 5, 2015, pp. 2044-2052.

[12] Prosini, P. P., “Modeling the voltage profile for LiFePO4,” *Journal of the Electrochemical Society*, Vol. 152, No. 10, 2005, pp. 1925-1929.

[13] Faruqui, A., Hledik, R., and Tsoukalis, J., “The power of dynamic pricing,” *The Electricity Journal*, Vol. 22, No. 3, 2009, pp. 42-56.

[14] Rastler, D. M. “Electricity energy storage technology options: a white paper primer on applications, costs and benefits,” *Electric Power Research Institute*, 2010.

[15] Allcott, H. “Real time pricing and electricity markets,” Harvard University (2009).

Chapter 5

Economics of Charging Services

5.1 Introduction

Investigating the economics underlying charging services furnishes a vision for designing and operating service stations. The benefits of such investigations are twofold. First, it provides a mechanism for decentralized control of the service requests, in which the behavior of the customers for requesting services and adjusting their request rates are governed primarily by incentives, rather than by strict rules and regulations. Second, optimizing service performance based on purely engineering performance figures of merit, such service delays and service blocking (denial) rates, while optimizing the economic performance, will add other wide-ranging measures such as service flexibility and adapting to the particular needs of the customers.

In this chapter, we discuss several economic issues that are instrumental to designing the control and pricing in electric vehicle charging stations. The discussions in this chapter, collectively, clarify the following four major issues that lie at the core of an economic vision for charging services.

- *Overprovision versus control:* While the demand for charging services is expected to grow into the future, there are many uncertainties about the extent of such demands. Any decision based on only predictive models can easily lead to underestimating or overestimating the demands. There are, in general, two remedies for addressing this. One solution entails overprovisioning according to which the networks are designed with simple structures but substantially large capacities based on the most aggressive predictions. Such solutions lead to overestimating the needs, but also they ensure that the networks are never congested. Another solution prescribes including less capacity, and subsequently, reducing the infrastructure cost, by adopting sophisticated control algorithms that coordinate service request rates and admission of the costumers to the charging stations. A holistic design for charging stations and customer service access will generally adopt both solutions and strikes a balance between the costs associated with overprovisioning and control. This topic is particularly important because the traditional power grid design has favored overprovisioning grid resources to meet the peak demand that occurs in a very small percentage of time at a very high cost.

- *Pricing as a control signal:* Dynamic pricing plays a pivotal role in controlling the rate of service requests. In principle, increasing the service rates reduces the service demand, reduces service congestion, and as a result reduces (or eliminates) service outages. In short time scales, dynamic pricing enables flexible control mechanisms, and in longer time scales, it allows for stabilizing the flow control and overall network operations.

- *Interconnected stations:* It is beneficial to the customers as well as the overall system performance that the charging stations are networked through an interconnected backbone. Such an interconnection facilitates information exchange and coordinated decision-making among all the stations. Specifically, creating larger networks from smaller ones allows for balancing out the level of services that different stations provide over time. This can avoid instances in which one station is being overcommitted by receiving frequent service requests, while other stations are being underutilized. This necessitates that stations constantly coordinate their signals and dynamic prices, which in turn, controls the service rates and the allocation of different customers to different stations.

Charging services can be deemed as economic goods that a network provides through its resources, among which the key one is the charging infrastructure. Next in this chapter we provide some of the basic economic concepts that are pivotal for formalizing the pricing dynamics in charging stations.

1. *Definitions:* In the subsequent sections and chapters we use *charge* as the amount that is billed for a service. By *service* we refer to an amount of monetary value associated with a unit of service, which is used for calculating the charge. Furthermore, *tariff* refers to the to the structure of prices and charges.

2. *Charging models:* In general, there are two types of tariffs based on which the service charges are calculated: flat-rate tariffs and usage-driven tariffs. In the former, the rates are flat and the customers are charged a fixed rate per service, while in the latter the price changes based on the priority of services, time of service, service duration, and other factors that capture the customer's preference and available network resources. Flat-rate charges, while having simple structures and being predictable to the customers, lack the versatility to cope with circumstances in which the network is congested with frequent service requests.

5.2 Charging for Services

Service charges and tariff structures depend on the demand (service request) and supply (charging capacity) dynamics and the market mechanisms (regulators) governing them. The demand for services is shaped by the value the customers place on their charging needs and the price they are willing to pay to obtain them. On the other hand, charging capacity is determined by how much the network suppliers can expect to charge for the services they provide as well as the infrastructure costs, which in turn, depend on network efficiency in properly striking a balance between controlling and overprovisioning the networks.

There are a number of rationales underlying the prices and tariff structures. The primary one stems from the intuitive fact that the prices are set based on the available technological capacities, the cost of energy supply, and the value that the customers place on the service that they receive. Based on this rationale, if the prices are set too high the stations might oversupply the network and there will be insufficient demand, while setting the prices too low leads to an insufficient incentive for the providers, and there is the likelihood that a portion of the demands will remain unsatisfied.

A second rationale pertains to the competition dynamics in circumstances in which there are multiple competing service providers. In such circumstances different charging service providers, on the one hand, need to balance their supply-demand dynamics and the associated economic dynamics, and on the other hand they need to protect their competitive advantages in the market.

The third rationale is related to the control actions that the providers wish to exert. Specifically, most of the congestion control and the ensuing stabilizing actions cannot be directly imposed on customers. As a remedy, providers leverage pricing dynamics to indirectly affect the rate and dynamics according to which the customers request service.

5.2.1 Consumer's Perspective

In this section we investigate the economic framework from the customers' perspectives. In this framework, consider a charging network of consisting of n customers, where each customer can request for k types of services (e.g., based on charging technology and duration). Denote the set of customers by $\mathcal{N} = \{1, \ldots, n\}$. Customer $i \in \mathcal{N}$ can request to receive x_j units of service $j \in \{1, \ldots, k\}$, resulting in the service vector $\boldsymbol{x} = [x_1, \ldots, x_k]$, for a payment of $p(\boldsymbol{x})$. Suppose that $p(\boldsymbol{x}) = \boldsymbol{p}^T \cdot \boldsymbol{x} = \sum_{j=1}^{k} p_j \cdot x_j$, for a given price vector $\boldsymbol{p}$. When the amount of the k services requested is unlimited (finite capacity constraints will be discussed in the following sections), costumer $i \in \mathcal{N}$ seeks to solve

$$\boldsymbol{x}^i(\boldsymbol{p}) = \arg\max_{\boldsymbol{x}} \left[u_i(\boldsymbol{x}) - \boldsymbol{p}^T \cdot \boldsymbol{x} \right] , \tag{5.1}$$

where u_i captures the utility of customer i, which can be interpreted as the price that customer i is willing to pay to receive the k services in quantities $\boldsymbol{x}$. It is often assumed that function u_i is strictly increasing and strictly concave in all its arguments. This ensures that the price decreases as the demand increases, and that there exists a unique maximizer for (5.1). These assumptions are typically made for designing the pricing dynamics. However, a combination of them can be relaxed when it suits the specifics of a setting.

The function that is maximized in (5.1) is referred to as the customer's *net benefit* or the customer's *surplus* i.e.,

$$\text{customer's benefit} = \max_{\boldsymbol{x}} \left[u_i(\boldsymbol{x}) - \boldsymbol{p}^T \cdot \boldsymbol{x} \right] . \tag{5.2}$$

This captures the net value that customers obtain as a function of $\boldsymbol{x}$ minus the amount that the customer pays for obtaining $\boldsymbol{x}$. Also the quantity $\boldsymbol{x}^i(\boldsymbol{p})$ given in (5.1) is referred to as the *demand function* for customer $i \in \mathcal{N}$.

Consider a single-customer network, who is deciding about the quantity to acquire from only one service j. If the customer's utility function u is concave and twice differentiable in x_j, then its net benefit is maximized where it is the stationary point with respect to x_j. Hence a customer's surplus who requests only for one type of service is given by

$$\text{customer's surplus at price } p_j = \int_0^{x_j(p_j)} p_j(\boldsymbol{x}) d\boldsymbol{x} - p_j x_j(p_j) .$$

We note that the model used in (5.1) relies on the assumption that all units of the same service have the same price. There exist, however, more general pricing mechanisms in which different service units are charged at different rates.

Concavity of the function u_i ensures that both functions $\boldsymbol{x}(\boldsymbol{p})$ and $p(\boldsymbol{x})$ are decreasing in their arguments, according to which the demand decreases as the price increases. Such a decrease in the rates is quantified by the index

$$\epsilon_j = \frac{\partial x_j(\mathbf{p})/\partial p_j}{x_j/p_j} , \tag{5.3}$$

which is referred to as the *price elasticity of demand*. Elasticity captures the change in the demand normalized by the rate of change in price. In certain circumstances, the demand for one service can affect the prices of other services. This occurs when services are substitutes (e.g., one service station provides charging services to replace a competing one) or complements (e.g., multiple charging stations coordinate their servicing schedules). In such cases, the notion of *cross elasticity of demand* is defined to capture the rate of change of the demand of a service normalized by the change in the price of another service. Specifically, for services i and j the cross elasticity ϵ_{ij} is defined as

$$\epsilon_{ij} = \frac{\partial x_i(\mathbf{p})/\partial p_j}{x_i/p_j} . \tag{5.4}$$

5.2.2 Service Provider's Perspective

Complementary to the discussion in the previous section, here, we consider a pricing and economic framework from the standpoint of the charging service providers. In this framework, consider a charging station (or a network of charging stations as one entity) that provides k different charging services. Define $\boldsymbol{y} = [y_1, \ldots, y_k]$ as the quantities of these k services. The quantities are bound to belong to known sets, which are shaped by the physical and capacity constraints. The *profit* of the provider is the difference between the revenue that is obtained by charging the customers, denoted by $r(\boldsymbol{y})$ and the cost of infrastructure and other resources consumed to provide the service, denoted by $c(\boldsymbol{y})$. Hence, profit maximization can be cast as

$$\text{profit} = \max_{\boldsymbol{y}} \left[r(\boldsymbol{y}) - c(\boldsymbol{y}) \right] . \tag{5.5}$$

There are three general strategies for optimizing the provider's benefits. In one extreme strategy the service provider is a *monopolist*, which is the sole service provider in an unregulated market. In this strategy, the service provider can freely set its prices, recognizing the inherent fact that increasing the prices can lead to a decrease in the customer demands. In another extreme strategy, the service provider is a small stakeholder compared to the pool of existing service providers. In such circumstances, the provider's decision influences the demand-response dynamics only marginally, and often the prices are set to change linearly with the service units. In the third strategy, the service provider can have a non-negligible partial influence over the prices, where it is one of the main providers and its control decisions affect other providers' decisions.

5.2.3 Socially Optimal Pricing

Constructed on the profit models developed for customers and service providers, next we provide a framework in which an aggregate of these individual profits are optimized, leading to a *socially optimal* tariff structure. In order to form the socially optimal structures, profits can be weighted differently, which allows for placing the desired emphasis on the contributions of different players to the aggregate profit. For instance, placing more attention on the customers' weights emphasizes the significance of the priorities for the customers' preferences. Under appropriate assumptions on the utility functions adopted for service providers and customers, maximizing the social profit leads to a pricing mechanism that the customers use to adjust their service request rates and durations. One significance of such socially optimal strategies is that they are amenable to distributed implementation, in which the customers' behaviors are coordinated via the pricing signals.

In order to formalize the aggregate profit and the associated framework, we consider a network of one charging station and a set of customers $\mathcal{N} = \{1, \ldots, n\}$. We denote the vector of quantities of k services requested by customer $i \in \mathcal{N}$ by $\boldsymbol{x}^i$. We denote the aggregate demand by $\boldsymbol{x}_T = \sum_{i=1}^{n} \boldsymbol{x}^i$. We define $c(\boldsymbol{x})$ as the cost of the charging stations for producing $\boldsymbol{x}$. The aggregate (social) profit is, therefore, the aggregate utility of the serviced consumed by all customers minus the cost of producing these services. Hence,

$$\text{aggregate profit } = \sum_{i=1}^{n} u_i(\boldsymbol{x}^i) - c(\boldsymbol{x}) . \tag{5.6}$$

Hence, obtaining the socially optimal pricing structure can be found as the solution to

$$\text{socially optimal profit} = \begin{cases} \text{maximize}_{\boldsymbol{x}} & \left[\sum_{i=1}^{n} u_i(\boldsymbol{x}^i) - c(\boldsymbol{x})\right] \\ & \\ \text{s.t.} & \boldsymbol{x} = \sum_{i=1}^{n} \boldsymbol{x}^i \end{cases} . \tag{5.7}$$

5.3 Other Pricing Constraints

The pricing model and the associated optimization problems described earlier rely on several simplistic assumptions that do not account for the delicate physical constraints. In this section, we review some of such constraints that entail dispensing with some of the simplifying assumptions.

5.3.1 Finite Capacity

Service providers have finite capacity, which enforces a hard constraint on the number of customers that can be served concurrently. The capacities for various charging types are set when the infrastructure is designed and built, and they remain fixed for a long period, and until the infrastructure can be upgraded. By invoking the finite-capacity constraints, we assume that the charging station can provide only C_j units of resource $j \in \{1, \ldots, k\}$ at any given time. This immediately leads to the following additional constraint

$$\sum_{i=1}^{n} x_j^i \leq C_j , \quad \forall j \in \{1, \ldots, k\} . \tag{5.8}$$

By including this additional into (5.7), the socially optimal profit can be found as the solution to

$$\text{socially optimal profit} = \begin{cases} \text{maximize}_{\boldsymbol{x}} & \left[\sum_{i=1}^{n} u_i(\boldsymbol{x}^i) - c(\boldsymbol{x})\right] \\ \text{s.t.} & \boldsymbol{x} = \sum_{i=1}^{n} \boldsymbol{x}^i \\ & \sum_{i=1}^{n} x_j^i \leq C_j \,, \quad \forall j \in \{1, \dots, k\} \end{cases} . \tag{5.9}$$

5.3.2 Auction-based Pricing

Finding the solution to the optimal pricing structure can sometimes involve performing an iterative algorithm (e.g., in auction-based pricing). In such circumstances, guaranteeing that the iterations converge to a socially optimal solution is of paramount significance. In such cases, when the demand curves are sensitive to the prices, any slight deviation from the true optimal price value can translate to a substantial deviation of the demands and customers behavior from their socially optimal values. This issue is also often referred to as the convergence of the *tatonnement* process.

5.3.3 Untruthful Behavior

In the previous discussions, we have assumed that the demand of any customer is negligible compared with the entire charging capacity of the network, and the aggregate demand that customers demand collectively. Nevertheless, in smaller networks, the charging demand of one user can affect the dynamics of the entire network, and in such circumstances, if a customer decides to be untruthful about its demands and charging needs, it can compromise network operations. The potential existence of such circumstances necessitates devising mechanisms that can make the operations robust against potentially untruthful reports by the customers.

5.4 Teletraffic Theory

The teletraffic theory aims to adopt theories from traffic theory, which primarily consists of statistical analysis and queuing theory, and apply them to analyzing telecommunication and data networks. Recent studies have been arguing some equivalence between data networks and power grids. Specifically, the grid serves multiplexed electricity demands of end-customers similarly to a data network that carries multiplexed traffic from data sources. Also, similar to Internet traffic, the demand for electricity is variable and exhibits burstiness over multiple time scales. This allows for leveraging well-studied concepts and theories from teletraffic theory to address certain stochastic and scheduling problems in power engineering (e.g., battery sizing in storage systems). The teletraffic theory has been developed over the past three decades to model and analyzes data networks (e.g., asynchronous transfer mode (ATM) networks) that guarantee certain levels of quality of service. These techniques can be used to size assets in the electrical grid. The study in [1] establishes an equivalence theorem that facilitates modeling a branch of a distribution network as a simple queueing system. The teletraffic theory, in principle, strives to characterize the interplay among system size (dimensions), consumers' quality of service requirements, and system traffic (demands).

5.5 Congestion and Its Effect on Pricing

In previous sections, we have assumed that the utility (profit) of each customer depends only on the amount of service received and the parameters associated with the service. This assumption, however, relies on the premise that the network is not experiencing any congestion. Specifically, once we know the consumers' demand functions, the systems' cost functions and their technology sets, the profit (consumers' profit, system's profit, or a social profit) can be optimized by using prices to allocate services to the consumers who value them most and to match the demand for services to supply. These assumptions are all true for services that have statically defined guarantees that may not vary during service.

In many realistic settings, nevertheless, it can often happen that due to the excessive demand that cannot be supplied by the system, certain consumers face service outage. Such circumstances, in particular, occur when the parameters of services (e.g., the rates) can change dynamically during the lifetime of a service, and there is no strict guarantee on minimum performance levels. Under these assumptions, consumers tend to request the highest level of service possible from the network.

The models in (5.1)-(5.2) do not capture the effects of arbitrary congestion levels if they do not implement measures to limit the maximum congestion levels. To address this, these models can be modified so that when there exists a common resource pool to be shared by consumers, any user i has a net benefit that is characterized by the amounts of service demanded by other users. Hence, the net benefit of user i is

$$u_i(\boldsymbol{x}^i, \boldsymbol{y}) - \boldsymbol{p}^T \cdot \boldsymbol{x}^i , \tag{5.10}$$

where we have

$$\boldsymbol{y} = \frac{1}{k} \sum_{i=1}^{n} \boldsymbol{x}^i ,$$

for some constant k that is controlled by the resource capacity. If user i requests a quantity of service that is small compared with the total requests of all users, then for different values of $\boldsymbol{x}^i$ the fluctuation of $\boldsymbol{y}$ will be insignificant, and the problem of maximizing (5.10) simplifies to that of maximizing (5.2), while $\boldsymbol{y}$ is treated as a fixed parameter. In this section, however, we suppose $\boldsymbol{y}$ is not fixed, and consider the problem of determining p so that when the market is in equilibrium we maximize some measure such as social welfare or the service provider's profit.

Next, we present a simple model for congestion pricing. We consider the case of a single service quantified in terms of a single dynamic parameter. We consider n consumers, each requesting different amounts of this service. The service provider can provide ℓ units of service at cost $c(\ell)$. By first considering a fixed ℓ optimizing the social welfare can be cast as

$$\max_{x_i \geq 0, \forall i} \sum_{j=1}^{n} u_j(x_j, y) - c(\ell) , \tag{5.11}$$

where $y = \frac{1}{\ell} \sum_{j=1}^{n} x_j$. The utility functions u_j are assumed to be concave in x_j and strictly decreasing and convex in y. Hence, the maximum occurs at the stationary point

$$\frac{\partial u_j(x_j, y)}{\partial x_j} + \frac{1}{\ell} \sum_{j=1}^{n} \frac{\partial u_j(x_j, y)}{\partial y} = 0, \quad \forall j \in \{1, \ldots, n\} . \tag{5.12}$$

Solving this set of equations results in socially optimal demands, denoted by $\{x_1^*, \dots, x_n^*\}$. Based on these optimal values, when user i is offered the price

$$p_E = -\frac{1}{\ell}\sum_{j=1}^{n}\frac{\partial u_j(x_j^*, y)}{\partial y}\,, \tag{5.13}$$

which is guaranteed to be positive since u_j is decreasing in y. Based on this price, consumer i seeks to optimize

$$\max_{x_i \geq 0}\ \left[u_j(x_j, y) - p_E x_i\right], \tag{5.14}$$

which leads to

$$\frac{\partial u_j}{\partial x_j} + \frac{1}{\ell}\frac{\partial u_j}{\partial y} - p_E = \frac{\partial u_j}{\partial x_j} + \frac{1}{\ell}\frac{\partial u_j}{\partial y} + \frac{1}{\ell}\sum_{j=1}^{n}\frac{\partial u_j(x_j^*, y)}{\partial y} = 0\,. \tag{5.15}$$

When n is sufficiently large we have

$$\left|\frac{1}{\ell}\frac{\partial u_j}{\partial y}\right| \ll \left|\frac{1}{\ell}\sum_{j=1}^{n}\frac{\partial u_j(x_j, y)}{\partial y}\right|\,. \tag{5.16}$$

Hence, the solution to (5.15) will be well-approximated by that of (5.12).

5.6 Network Externalities

It is often assumed that consumers' profits (utilities) depend on the amount of service they receive. This assumption, however, might break down as more consumers use the services. For instance, consider a network of n consumers in which consumer i is willing to pay $u_i(j) = ji$ for a service unit when j consumers will be also receiving the service. Hence, if consumer i believes that no other consumer will receive this service, it will value this service at 0.

Analyzing the profits under such models suggests that service markets with strong network effects may remain limited and never really reach the socially desirable points of high penetration. This type of market crash can arise unless certain incentives move the operational point of the market to the desired optimal ones. Such failures may occur either because enough customers have initially high expectations of the eventual market size (due to, e.g., successful marketing), or because a social entity subsidizes the cost of the service, resulting in a lower posted price. These conditions are frequently encountered in various service markets, in which wide penetration of the services forces low prices for access services. Nevertheless, prices will be low for access once sufficient demand for the services attracts more competition in the provision of such services and motivates the development and deployment of more cost-effective access technologies.

5.7 Demand Forecasting

Reliable load prediction in electricity grids is instrumental to efficient operation, control, and management of electric utilities. Load prediction is primarily responsible for delineating the amount of energy to be delivered at different geographical locations over different time horizons. Broadly, there are two categories of load prediction coping with prediction over short-term intervals (e.g., hourly, daily, weekly, and monthly) and long-term intervals (e.g., annually).

Long-term load prediction deals with forecasting the demands on an annual basis and is governed by multiple factors, including the spatial and temporal expansions of the networks, end-user appliances, urbanization dynamics, customer behavior, and population variations, to name a few. Depending on the nature of future demands, there are two general scenarios for long-term load prediction. In one scenario future demands are dominantly specified by spatial and temporal load growth and historical data is of secondary significance [2, 3]. In the other scenario, historical patterns of consumption and their pertinent underlying statistical models exhibit the dominant role in forecasting future demands [4–6]. The latter scenario is more apt for the networks that remain relatively static in scale and operation over long periods, in which future consumption will be strongly correlated with historical data.

Short-term load prediction, on the other hand, is essential for monitoring and controlling the power flow in the network on sub-annual scales (e.g., hourly, daily, weekly, and monthly). A wide variety of short-term prediction models exists in the literature that differ in complexity, underlying statistical assumptions, available historical data, and prediction procedures. The major existing approaches include the following methods: multiple regression [7, 8], iterative re-weighted least-squares [9], stochastic time-series [10, 11], autoregression models based on genetic algorithms [12, 13], neural networks [14, 15], and knowledge-based approaches.

Depending on the network dynamics, these methods offer different advantages and, consequently, selecting an appropriate method for a network is strongly coupled with the model and information available about the network.

Selecting an appropriate prediction model can often be difficult, and it becomes even more severe as the future grids are anticipated to be complex networked platforms in which high-dimensional data is routinely generated, exchanged, and processed. Such complexity stems from the distributed nature and size of the grid, institutional deregulations that allow integrating private and small producers into the grid, and the abundant presence of intermittent energy sources, such as renewables. Hence, the future smart grid will not only be complex in terms of its constituents, but also will be unpredictable in its nature.

5.8 Chapter Summary

In this chapter we investigated some of the major issues that are critical and instrumental to designing economic frameworks underlying charging services. Such mechanisms are necessary in order to exert sufficient leverages for properly provisioning and controlling the rates at which resources are being consumed by a network of consumers. We specifically discussed the balance between overprovisioning resources and controlling consumers' resource access patterns, and provided an overview of how pricing can act as a control signal. Furthermore, we provided an overview of optimizing resource allocation from the perspectives of consumers and service providers, and some of the challenges addressed that will require abstracting and incorporating them into the resource allocation optimization process.

References

[1] Ardakanian, O., Keshav, S., and Rosenberg, C. , "On the use of teletraffic theory in power distribution systems," in *Proc. Third International Conference on Future Energy Systems: Where Energy, Computing and Communication Meet*, Madrid, Spain, Jun. 2012.

[2] Brown, R., Hanson, A., and Hagan, D."Long range spatial load forecasting using non-uniform areas," in *Proc. IEEE Transmission and Distribution Conf.*, Raleigh, NC, Apr. 1999.

[3] Willis, H., Finley, L., and Buri, M., "Forecasting electric demand of distribution system planning in rural and sparsely populated regions," *IEEE Trans. Power Syst.*, Vol. 10, No. 4, pp. 2008–2013, Nov. 1995.

[4] Fu, C. W. and Nguyen, T. T. "Models for long-term energy forecasting," in *Proc. IEEE Power Eng. Soc. General Meeting (PES)*, Crawley, Australia, July 2003.

[5] Kandil, M., El-Debeiky, S., and Hasanien, N. E., “Long-term load forecasting for fast developing utility using a knowledge-based expert system,” *IEEE Trans. Power Syst.*, Vol. 17, No. 2, pp. 491–496, May 2002.

[6] Xia, C., Wang, J., and McMenemy, K., “Short, medium and long term load forecasting model and virtual load forecaster based on radial basis function neural networks,” *Int. J. Elec. Power*, Vol. 32, No. 7, pp. 743–750, Sept. 2010.

[7] Moghram, I., and Rahman, S.,“Analysis and evaluation of five short-term load forecasting techniques,” *IEEE Trans. Power Syst.*, Vol. 4, No. 4, pp. 1484–1491, Nov. 1989.

[8] Papalexopoulos, A. and Hesterberg, T., “A regression-based approach to short-term system load forecasting,” *IEEE Trans. Power Syst.*, Vol. 5, No. 4, pp. 1535–1547, Nov. 1990.

[9] Mbamalu, G., “Load forecasting via suboptimal seasonal autoregressive models and iteratively reweighted least squares estimation,” *IEEE Trans. Power Syst.*, Vol. 8, No. 1, pp. 343–348, Feb. 1993.

[10] El-Keib, A. A., Ma, X., and Ma, H.,“Advancement of statistical based modeling techniques for short-term load forecasting,” *Electr. Pow. Syst. Res.*, Vol. 35, No. 1, pp. 51–58, Oct. 1995.

[11] Huang, S.,“Short-term load forecasting using threshold autoregressive models,” *IET Gener. Transm. Distrib*, Vol. 144, No. 5, pp. 477– 481, Sep. 1997.

[12] Yang, H.-T. ,“Identification of armax model for short term load forecasting: an evolutionary programming approach,” *IEEE Trans. Power Syst.*, Vol. 11, No. 1, pp. 403–408, Feb. 1996.

[13] Yang, H. T., and Huang, C. M., “A new short-term load forecasting approach using self-organizing fuzzy armax models,” *IEEE Trans. Power Syst.*, Vol. 13, No. 1, pp. 217–225, Feb. 1998.

[14] Sinha, A. and Mondal, J.,“Dynamic state estimator using ann based bus load prediction,” *IEEE Trans. Power Syst.*, Vol. 14, No. 4, pp. 1219–1225, Nov. 1999.

[15] Hippert, H. and Pedreira, C., and Souza, R., “Neural networks for short-term load forecasting: a review and evaluation,” *IEEE Trans. Power Syst.*, Vol. 16, No. 1, pp. 44–55, Feb. 2001.

[16] Kekatos, V., and Giannakis, G., “Distributed robust power system state estimation,” *IEEE Trans. Power Syst.*, Vol. 28, No. 2, pp. 1617–1626, May 2013.

[17] Zhao, L. and Abur, A.,“Multi area state estimation using synchronized phasor measurements,” *IEEE Trans. Power Syst.*, Vol. 20, No. 2, pp. 611–617, May 2005.

Chapter 6

Capacity Planning for Large-Scale PEV Charging Stations

6.1 Introduction

As discussed in previous chapters, the proliferation of PEVs requires the deployment of large-scale public charging stations. Such stations are installed in parking lots, which can be deployed in hospitals, shopping malls, airports, and workplace campuses. Such facilities are essential, and possibly the single option based on the current technology, for the fleet driven in populated regions. These facilities will also provide complementary service to expand the electric range for PEV drivers who have access to garage charging. Effective and efficient deployment of such charging facilities, however, involves challenges beyond only upgrading the existing stations with the necessary chargers. As discussed in Chapters 1 and 2, a large-scale PEV charging station can be equipped with either level 2 single/three phase, or fast DC charging technologies [1]. Incorporating these technologies translates to substantial phase additions into the existing power grid.

For example, during a 30-minute period, level 2 single-phase (3.3kW), level 2 three-phase (6.6 kW), and DC fast chargers (50 kW) can enable a typical EV to drive 5.5, 11, and 83.4 miles, respectively. On the other hand, such infrastructure is crucial to PEV adoption, since according to a related survey [2], even with the present adoption level, 30% of vehicles use public charging regularly and 40% of PEV drivers travel further than their all-electric-range and, as a consequence, they need to use public charging to support their daily trips.

As discussed in Chapter 3, stochastic PEV demand exerted on the power grid threatens the stability of the power grid operations, especially the distribution grid. This necessitates a thorough and careful design for such facilities. In this chapter we address some of the challenges associated with such designs in the context of large-scale charging stations that are equipped with energy storage devices. Specifically, the following two main problems are addressed in this chapter:

- We introduce a large charging station architecture, formalize a probabilistic framework to capture its operational characteristics, and assess its performance (defined as the percentage of served PEVs). The charging facility is equipped with a local energy storage device, which aids smoothing stochastic PEV demands.
- As summarized in Chapter 2, by considering their high acquisition, operation, and maintenance costs, we present a stochastic framework for analyzing the optimal size of current energy storage technologies. Optimal sizing of energy storage systems (ESSs) plays a critical role in balancing the efficiency and reliability of charging stations. In principle, overprovisioning ESS sizes lead to under-utilizing costly assets and underprovisioning them taxes operation lifetime.

The design of charging facilities is ultimately linked to the current power grid operations. PEVs represent sizable, *mobile* electric loads. Level 1 charging represents a load comparable to a household, while level 2 charging represents a load twice as large as a household. Thus, a large number of PEVs that are geographically concentrated impose considerable strains not only on power generation, but also on the grid distribution system. Charging station designs that do not stress the power grid and eliminate the need for adding significant extra generation, therefore, have paramount significance. The key components of our design mechanism leveraged to address these challenges are as follows:

- First, charging stations draw constant power from the grid for a predefined period of time and charging stations of any significant size represent commercial size loads. Hence, it is beneficial for station operators to draw long-term contracts with the utility by which a minimum power level is committed to in return for a lower price. This reduces the uncertainties associated with PEV demands and enables the utility to forecast its demand more accurately. Furthermore, station operators benefit from a lower price, as analyzed in [9]. As a result, since

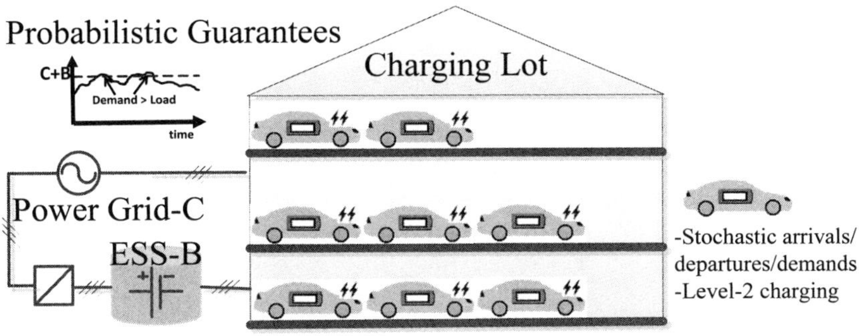

Figure 6.1 Charging station model.

such contracts lead to lower spot prices and more efficient market equilibria, implementing them benefits the charging stations.

- Second, local energy storage is employed to buffer and aid stochastic fluctuations of the PEV demands. During peak hours, stored energy can be used to serve more PEVs. Conversely, when the power that the station can draw from the grid is not fully utilized, the excess power can recharge the storage device.
- Third, the quality of service (QoS) metric employed is the long-term blocking probability of incoming PEVs. This metric is used to size the energy storage device.
- Finally, charging stations can provide service to multiple PEV classes at different charging rates. This allows the stations to accommodate PEVs with different charging needs and preferences, as well as PEVs with different technological constraints. The system model is depicted in Figure 6.1

6.2 System Description

We consider a large-scale charging station in which the demands of N PEVs are accommodated by the power grid along with a *shared* energy storage system unit. The grid capacity at time t is denoted by C_t and ESS depletion level is $S(t)$, for $t \in \mathbb{R}^+$. Furthermore, the energy storage unit has the following parameters [4, 5]:

- *Energy rating* or the size of the storage is represented by B (kWh).
- *Power rating* is the rate at which storage can be charged or discharged. The charging rate is assumed to be $P_c \leq C_t, \forall t$, and discharge rating P_d is related to the energy rating B since $B = P_d \times$ (desired support duration). Desired support duration is usually equivalent to the duration of the peak hour.
- *Efficiency* of a charge-discharge cycle is denoted by $\eta \in [0, 1]$, which shows the fraction of the stored energy or the fraction of energy that is transferred after losses are extracted.
- *Dissipation losses* represent a small percentage of losses that occur due to leakage. For simplicity in notations, dissipation losses are ignored.

Furthermore, since PEVs do not necessarily have identical demands, we consider K PEV classes, which are distinguished by the amount of their demands, which are, in turn, represented by $\{R_k\}$ as the energy demand per time unit for PEV type $k \in \{1, \ldots, K\}$. Let N_k denote the number of PEVs of type k and the vector $\boldsymbol{N}$ represent the number of PEVs of each type; that is, $\boldsymbol{N} = (N_1, \cdots, N_K)$. As established in [1, 6], the consumption pattern of each PEV can be well-represented by a two-state On/Off process. We define the binary variable s_t^{ik} to represent the state of PEV i of type k at time t such that

$$s_t^{ik} = \begin{cases} 1 & \text{PEV } i \text{ is On} \\ 0 & \text{PEV } i \text{ is Off} \end{cases} . \tag{6.1}$$

When a PEV is in the On state, it initiates an energy demand. The duration of demand is modeled statistically, which is adopted to capture the variety in the types of PEVs' demands (mainstream PEV packs are given in Chapter 1). Since different customer types do not necessarily charge their batteries at the same rate, we denote the *average* charging duration of customers of type k by $1/\mu_k$. It is noteworthy that from the station operator's standpoint, the information required to characterize customers' profiles are the average service duration and the charging rates (depicted in Figure 6.2). These parameters enable modeling customers' behaviors with fine granularity. Furthermore, we assume that the requests, which are transitions from Off to On, are generated randomly and according to a Poisson process with parameter λ_k. Hence, for each PEV i of type k at any time t we have

$$\mathbb{P}(s_t^{ik} = 1) = \frac{\lambda_k}{\lambda_k + \mu_k} . \tag{6.2}$$

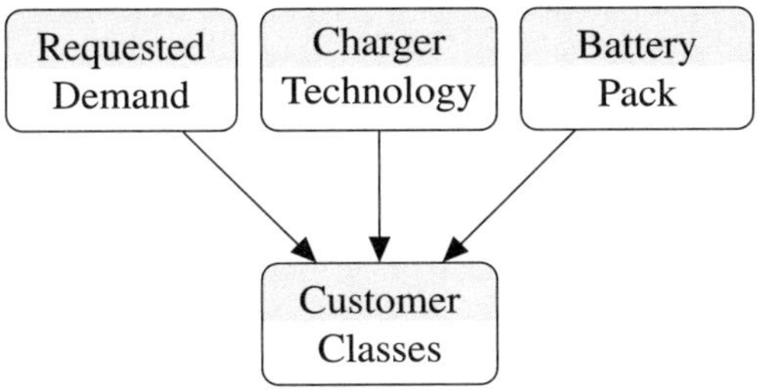

Figure 6.2 Customer classes are determined by PEV battery pack, amount of requested demand, and the charger technology.

In order to formalize the dynamics of ESS, we define $L_{ik}(t)$ as the charge request of PEV i of type k. Then, there are exactly three cases that define the rate of change in the storage unit: (1) the storage can be in the fully charged state, and the aggregate demand is less than the grid power C_t; (2) the storage can be completely discharged, and the aggregate demand is less than the grid power C_t; and (3) any partially charged state with any PEV demand. Therefore, for the rate of change in the storage level of the ESS we have

$$\frac{dS(t)}{dt} = \begin{cases} 0\,, & \text{if } S(t) = B \ \& \\ & \sum_{k=1}^{K}\sum_{i=1}^{N_k} \eta L_{ik}(t) < C_t \\ 0\,, & \text{if } S(t) = 0 \ \& \\ & \sum_{k=1}^{K}\sum_{i=1}^{N_k} \eta L_{ik}(t) > C_t \\ \eta(C_t - \sum_k \sum_i L_{ik}(t))\,, & \text{otherwise} \end{cases} . \tag{6.3}$$

Due to stochasticities involved (in consumption and generation), by selecting any energy storage capacity B, only probabilistic guarantees can be provided for the reliability of the power grid, and there exists a possibility of a blackout, which occurs when available resources drop below the aggregate demands by the PEVs. Note that $S(t)$ expresses the energy level that the storage device needs to feed into the grid to avoid an outage, which occurs when the necessary load from the storage unit exceeds the maximum available B. Hence, we define the ε-outage storage size, indicated by $B(\varepsilon)$, as the smallest selection of B corresponding to which the likelihood of blackout does not exceed $\varepsilon \in (0, 1)$, i.e.,

$$B(\varepsilon) = \begin{cases} \text{minimize} & B \\ \\ \text{subject to} & \mathbb{P}(S_t \geq B) \leq \varepsilon \end{cases} . \tag{6.4}$$

Our goal is to determine the ε-outage storage capacity $B(\varepsilon)$ based on grid capacity C_t, the number of PEVs N, and their associated consumption dynamics. For the simplicity of mathematical expressions, we scale the storage parameters and instead of B/η we redefine B as the maximum amount of energy that can be stored, and similarly P_d and P_c as the actual power ratings. We will start our analysis by deriving $B(\varepsilon)$ for a single class case, which will be leveraged later for analysis in the multiclass case.

6.3 Storage Capacity Analysis for Single-class PEVs

6.3.1 Energy Storage Access Dynamics

When the power grid can charge all the PEV batteries, there will be no PEV served by the energy storage device. On the other hand, when the grid capacity falls below the aggregate demand, the PEVs rely on getting access to the storage unit. Since the requests of the PEVs arrive randomly, the number of PEVs accessing the storage unit will also vary randomly.

Since we have N *independent* PEVs, each with a two-state model, by considering their underlying arrival and consumption processes, the composite model counting the number of PEVs accessing the storage unit at a given time can be modeled as a *continuous*-time birth-death process. Specifically, this process consists of $(N + 1)$ states, in which state $n \in \{0, \ldots, N\}$ models n PEVs being active and accessing the energy storage unit; that is,,

$$\text{state at time } t \text{ is } n \text{ if } \sum_{i=1}^{N} s_t^i = n , \tag{6.5}$$

and drawing nR_p units of power from the storage unit. As depicted in Figure 6.3, the transition rate from state n to state $n+1$ is $(N - n)\lambda$ and, conversely, the transition from state $n + 1$ to state n is $(n + 1)\mu$. Hence, for the associated infinitesimal

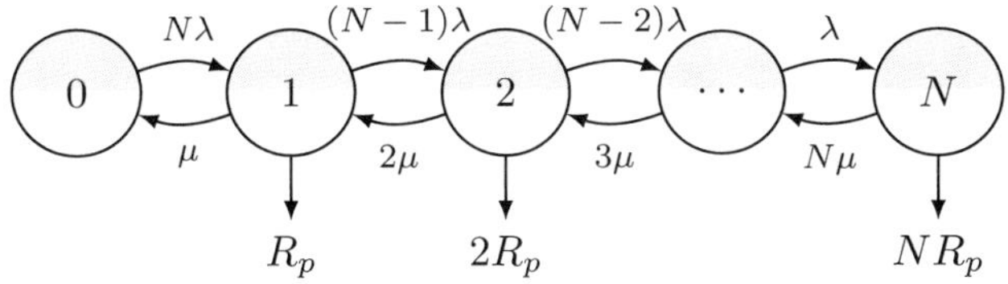

Figure 6.3 Composite model for N independent PEVs for single PEV type (k=1). Each PEV becomes active (On) at rate λ and becomes inactive (Off) at rate μ. The aggregate demand depends on the active number of PEVs.

generator matrix M, in which the row elements sum to zero, for $i, n \in \{0, \ldots, N\}$ we have

$$M[i,n] = \begin{cases} -((N-i)\lambda + i\mu) & n = i \\ i\mu & n = i-1 \ \& \ i > 0 \\ (N-i)\lambda & n = i+1 \ \& \ i < N \\ 0 & \text{otherwise} \end{cases} . \tag{6.6}$$

By denoting the stationary probabilities of state $n \in \{0, \ldots, N\}$ by π_n, and accordingly defining $\boldsymbol{\pi} = [\pi_0, \pi_1, ..., \pi_N]$, these stationary probability values satisfy $\boldsymbol{\pi} M = \mathbf{0}$.

6.3.2 Analyzing Distributions

In the next step, we analyze the statistical behavior of the ESS charge level. Specifically, we define $F_i(t,x)$ as the cumulative distribution function (CDF) of the ESS charge level when $i \in \{0, \ldots, N\}$ PEVs are depleting the storage unit at time t,

$$F_i(t,x) \;=\; \mathbb{P}\left(S(t) \le x \quad \text{and} \quad \sum_{j=1}^{N} s_t^j = i \right) . \tag{6.7}$$

Consequently, we define the vector of cdfs as

$$\boldsymbol{F}(t,x) \triangleq [F_0(t,x) \ , \ F_1(t,x) \ , \ ... \ , \ F_N(t,x)] \ . \tag{6.8}$$

Based on these definitions, the following lemma provides a differential equation that admits the CDF vector as its solution and is instrumental to evaluating the probability of interruption events; that is, $\mathbb{P}(\sum_{i=}^{N} L_i(t) > C_t + B)$.

Lemma 1. The CDF vector $\boldsymbol{F}(t, x)$ satisfies

$$\frac{d\boldsymbol{F}(t,x)}{dx} \cdot D = \boldsymbol{F}(t,x) \cdot M \,, \tag{6.9}$$

where D is a diagonal matrix defined as

$$D \triangleq \text{diag}\left[-C_t\mu \,,\ (1 - C_t)\mu \,,\ \dots \,, (N - C_t)\mu\right] \,, \tag{6.10}$$

and matrix M is defined in (6.6).

Proof. In order to compute the probability density functions, we find the expansion of $F_i(t, x)$ for an incremental change Δt in t; that is, $F_i(t+\Delta t, x)$. Note that during incremental time Δt, three elementary events can occur:

1. One inactive PEV might become active (i.e., i increases to $i + 1$);
2. One active PEV becomes inactive, (i reduces to $i - 1$);
3. The number of active PEVs remains unchanged.

In other words, case 1 represents the event that a new PEV arrives, while case 2 represents a service completion event, and case 3 shows the unchanged situation. Since the periods of arrival and departure of PEVs are exponentially distributed, corresponding to these events, CDF $F_i(t, x)$ can be expanded as

$$\begin{aligned} &F_i(t + \Delta t, x) \\ &\quad = \underbrace{[N - (i - 1)] \cdot (\lambda \Delta t) \cdot F_{i-1}(t, x)}_{\text{one PEV added}} \\ &\quad + \underbrace{[i + 1] \cdot (\mu \Delta t) \cdot F_{i+1}(t, x)}_{\text{one PEV removed}} \\ &\quad + \underbrace{[1 - ((N - i)\lambda + i\mu)\, \Delta t] \cdot F_i(t, x - (i - C_t) \cdot \mu \Delta t)}_{\text{no change}} \\ &\quad + o\left((\Delta t)^2\right) , \end{aligned} \tag{6.11}$$

where $o\left((\Delta t)^2\right)$ represents the probabilities of the compound events and tends to zero more rapidly than $(\Delta t)^2$ (and Δt) as $\Delta t \to 0$. Next, by passing the limit

$$\lim_{\Delta t \to 0} \frac{F_i(t+\Delta t, x)}{\Delta t}$$

it can be readily verified that (6.11) simplifies to

$$\begin{aligned}\frac{\partial F_i(x,t)}{\partial t} &= [N-(i-1)] \cdot (\lambda) \cdot F_{i-1}(t,x) \\ &+ [i+1] \cdot (\mu) \cdot F_{i+1}(t,x) \\ &- [(N-i)\lambda + i\mu] \cdot F_i(t,x) \\ &- (i - C_t) \cdot (\mu) \cdot \frac{\partial F_i(t,x)}{\partial x} \ , \end{aligned} \tag{6.12}$$

where we have defined $F_{-1}(t,x) = F_{N+1}(t,x) = 0$. Recall that the main objective is to compute the ESS size that will operate over a long time period. Hence, it is further assumed that steady state condition holds; that is $\partial F_i(x,t)/\partial t = 0$. Hence, (6.12) can be rewritten as

$$\begin{aligned}(i - C_t) \cdot (\mu) \cdot \frac{\partial F_i(t,x)}{\partial x} &= [N-(i-1)] \cdot (\lambda) \cdot F_{i-1}(t,x) \\ &+ [i+1] \cdot (\mu) \cdot F_{i+1}(t,x) \\ &- [(N-i)\lambda + i\mu] \cdot F_i(t,x) \ . \end{aligned} \tag{6.13}$$

By concatenating all the equations (6.13) for all $i \in \{0, \ldots, N\}$ we obtain the compact form

$$\frac{d\boldsymbol{F}(t,x)}{dx} \cdot D = \boldsymbol{F}(t,x) \cdot M \ , \tag{6.14}$$

which is the desired result. ■

The solution of the first-order differential equation given in (6.14) can be expressed as a sum of exponential terms. The general solution requires computing $(N+1)$ eigenvalues of the matrix MD^{-1} and the general solution is expressed as [7]

$$\boldsymbol{F}(t,x) = \sum_{i=0}^{N} \alpha_i \ \boldsymbol{\phi}_i \ \exp(z_i x) \ , \tag{6.15}$$

where z_i is the ith eigenvalue of MD^{-1} with the associated eigenvector ϕ_i that satisfies $z_i\phi_i D = \phi_i M$. The coefficients $\{\alpha_0, \ldots, \alpha_N\}$ are determined by the boundary conditions; for example, $F_i(t, 0) = 0$ and $F_i(t, \infty) = 1$.

In order to compute the probability distribution in (6.15), we need to determine the eigenvalues of MD^{-1}, the eigenvectors ϕ_i, and coefficients α_i. We notice that, since $x \geq 0$ and $F_j(x)$ is upper bounded by 1, all of the positive eigenvalues and the corresponding α_i must be set to zero, hence this reduces the computational complexity and (6.15) simplifies to:

$$\boldsymbol{F}(t, x) = \sum_{i:\mathrm{Re}[z_i \leq 0]} \alpha_i \, \boldsymbol{\phi}_i \, \exp(z_i x) \, . \tag{6.16}$$

We further notice that from $z_i\boldsymbol{\phi_i}D = \boldsymbol{\phi_i}M$, one of the eigenvalues must be zero. Then by setting $z_0 = 0$, the corresponding eigenvector can be computed from $\boldsymbol{\phi}_0 M = \mathbf{0}$. Recall from the previous discussion that the steady state probability distribution $\boldsymbol{\pi}$ of the $(N + 1)$ state Markov chain can also be computed from the same equation; that is $\boldsymbol{\pi} M = \mathbf{0}$. Since the eigenvector $\boldsymbol{\phi}_0$ is known and one of the eigenvalues is $z_0 = 0$, we can write $\boldsymbol{\phi}_0 = \boldsymbol{\pi}$. Therefore, (6.16) further simplifies to [8]:

$$\boldsymbol{F}(t, x) = \boldsymbol{\pi} + \sum_{i:Re[z_i < 0]} \alpha_i \, \boldsymbol{\phi}_i \, \exp(z_i x) \, . \tag{6.17}$$

6.3.3 Single-charger Technology with One PEV

In order to establish how to compute the desired ε-outage storage capacity $B(\varepsilon)$, by leveraging the CDF vector found in (6.17), we start with a simple network with a single PEV ($N = 1$). The intuitions obtained can be leveraged to generalize the approach for networks with any arbitrary size N. When $N = 1$ the infinitesimal generator matrix M defined in (6.6) is

$$M = \begin{bmatrix} -\lambda & \lambda \\ \mu & -\mu \end{bmatrix} . \tag{6.18}$$

For finding the expansion of $\boldsymbol{F}(t, x)$ as given in (6.15), we need to find z_0 and z_1 as the eigenvalues of MD^{-1}, where D is defined in (6.10). Specifically, based on

(6.10) we find that

$$MD^{-1} = \begin{bmatrix} \frac{1}{C_t} \cdot \frac{\lambda}{\mu} & -\frac{1}{1-C_t} \cdot \frac{\lambda}{\mu} \\ -\frac{1}{C_t} & -\frac{1}{1-C_t} \end{bmatrix} . \tag{6.19}$$

Hence, the eigenvalues are

$$z_0 = 0 \quad \text{and} \quad z_1 = \frac{\chi}{C_t} - \frac{1}{1 - C_t} , \tag{6.20}$$

where we have defined $\chi \triangleq \frac{\lambda}{\mu}$. It can be readily verified that the eigenvector associated with z_1 is $\boldsymbol{\phi}_1 = [1 - C_t \ , \ C_t]$. Therefore, according to (6.17) we have

$$\boldsymbol{F}(t, x) = \boldsymbol{\pi} + \alpha_1 \ \boldsymbol{\phi}_1 \ \exp(z_1 x) . \tag{6.21}$$

Finally, by finding the coefficient α_1 we can fully characterize $\boldsymbol{F}(t, x)$. This can be facilitated by leveraging the boundary condition $F_1(t, 0) = 0$, which yields

$$F_1(t, 0) \ = \ \pi_1 \ + \ \alpha_1 \ C_t \ = \ 0 , \tag{6.22}$$

where we have that $\pi_1 = \frac{\lambda}{\lambda+\mu}$. Therefore

$$\alpha_1 \ = \ -\frac{\chi}{C_t(1 + \chi)} , \tag{6.23}$$

which subsequently fully characterizes both cdfs $F_0(t, x)$ and $F_1(t, x)$ according to

$$\begin{aligned} F_0(t, x) \ &= \ \pi_0 + \alpha_1(1 - C_t) \exp(z_1 x) , \\ \text{and} \quad F_1(t, x) \ &= \ \pi_1 + \alpha_1 C_t \exp(z_1 x) . \end{aligned}$$

As a result, by recalling the definition of $F_i(t, x)$ in (6.7), the probability that the storage level S_t falls below a target level x is given by

$$\mathbb{P}(S_t \leq x) \ = \ F_0(x) + F_1(x) = 1 + \alpha_1 \exp(z_1 x) . \tag{6.24}$$

Given this closed-form characterization for the distribution of S_t, we can now evaluate the probability term

$$\mathbb{P}(S_t > B) , \tag{6.25}$$

which is the main constraint in the energy storage sizing problem formalized in (6.4). Specifically, for any instantaneous realization of C_t denoted by c we have

$$\begin{aligned}\mathbb{P}(S_t > B) &= \int_{C_t} \mathbb{P}(S_t > B \mid C_t = c)\; f_{C_t}(c)\; dc \\ &= -\int_{C_t} \alpha_1 \exp(z_1 B)\; f_{C_t}(c)\; dc \\ &= \int_{C_t} \frac{\chi}{c(1+\chi)} \exp\left(\frac{B\chi}{c} - \frac{B}{1-c}\right) f_{C_t}(c)\; dc\,.\end{aligned}$$

Therefore, by noting that $z_1 = \frac{\chi}{c} - \frac{1}{1-c}$ is negative, the term $\mathbb{P}(S_t > B)$ becomes strictly decreasing in B. Hence, the smallest storage capacity B that satisfies the stochastic guarantee $\mathbb{P}(S_t > B) \leq \varepsilon$ has a unique solution corresponding to which this constraint holds with equality. In the simplest settings, in which grid capacity C_t is constant c, we find

$$B(\varepsilon) = \frac{c(1-c)}{\chi - \chi c - c} \cdot \log \frac{\varepsilon c(1+\chi)}{\chi}\,. \tag{6.26}$$

6.3.4 Single-charger Technology with Multiple PEVs

In this section we provide a closed form for the probability term $\mathbb{P}(S_t \leq x)$ for arbitrary values of N, which we denote by $F_N(x)$. Computing all $F_N(x)$ terms through computing its constituent terms $F_i(t, x)$, particularly since N grows, it becomes computationally expensive and possibly prohibitive as it involves computing the eigenvalues and eigenvectors of MD^{-1}. By capitalizing on the observation that for a large number of PEVs (i.e., $N \gg 1$) the largest eigenvalues are the main contributors to the probability distribution, the asymptotic expression for $F_N(x)$ is given by [9]

$$\begin{aligned}F_N(x) = \frac{1}{2}&\sqrt{\frac{u}{\pi f(\varsigma)(\varsigma + \lambda(1-\varsigma))N}} \\ &\times \exp(-N\varphi(\varsigma) - g(\varsigma)x) \\ &\times \exp(-2\sqrt{\{f(\varsigma)(\varsigma + \lambda(1-\varsigma))Nx\}})\end{aligned} \tag{6.27}$$

where we have defined

$$f(\varsigma) \triangleq \log\left(\frac{\varsigma}{\lambda(1-\varsigma)}\right) - 2\frac{\varsigma(1+\lambda)-\lambda}{\varsigma+\lambda(1-\varsigma)},$$

$$u \triangleq \frac{\varsigma(1+\lambda)-\lambda}{\varsigma(1-\lambda)},$$

$$\varphi(\varsigma) \triangleq \varsigma\log(\varsigma) + (1-\varsigma)\log(1-\varsigma) - \varsigma\log(\varsigma) + \log(1+\lambda),$$

$$g(\varsigma) \triangleq z + 0.5\,(\varsigma+\lambda(1-\varsigma))\,\frac{\psi(1-\varsigma)}{f(\varsigma)},$$

$$z \triangleq (1-\lambda) + \frac{\lambda(1-2\varsigma)}{(\varsigma+\lambda(1-\varsigma))},$$

$$\text{and}\quad \psi \triangleq \frac{(2\varsigma-1)(\varsigma(1+\lambda)-\lambda)^3}{\varsigma(1-\varsigma)^2(\varsigma+\lambda(1-\varsigma))^3}.$$

In this set of equations, time is measured in units of a single average On time $(1/\mu)$. Furthermore, κ and ς are defined as the ESS per PEV (B/N) and the grid power allocated per source, respectively. Furthermore, we denote the variable υ as the power above the mean demand allocated per PEV as $\upsilon = \varsigma - \frac{\lambda}{1+\lambda}$.

6.4 Storage Capacity Analysis for Multiclass PEVs

In this section, we consider the case where the storage unit serves more than one PEV type (e.g., level 1 and level 2). We start the analysis by noting that the continuous-time Markov chain model presented in Figure 6.3 becomes a K dimensional Markov process. Recall that total number of PEVs of each class is represented by $\boldsymbol{N} = (N_1, \cdots, N_K)$ and we let $\boldsymbol{n}(t) = (n_1, \cdots, n_K)$ denote the number of sources of type k that are in the On mode at time t. Furthermore, $\boldsymbol{n}(t)$ represents the state space of the Markov process and similar to the single-class case given in (6.6), the transitions can occur between neighboring states, and the transition matrix $\bar{M}$ is given by

$$\bar{M}(\boldsymbol{n}, \Delta_k^+(\boldsymbol{n})) = (N_k - n_k)\lambda_k, \tag{6.28}$$

$$\bar{M}(\boldsymbol{n}, \Delta_k^-(\boldsymbol{n})) = n_k\mu_k, \tag{6.29}$$

$$\bar{M}(\boldsymbol{n}, \boldsymbol{n}') = 0, \text{otherwise}. \tag{6.30}$$

where we have defined

$$\Delta_k^+(n_1, \cdots, n_k, \cdots, n_K) = (n_1, \cdots, n_k + 1, \cdots, n_K) , \tag{6.31}$$

$$\text{and} \quad \Delta_k^-(n_1, \cdots, n_k, \cdots, n_K) = (n_1, \cdots, n_k - 1, \cdots, n_K) . \tag{6.32}$$

We provide a toy example to clarify this. Recall that in the single-class case, the transitions can only occur between two neighboring states. In the multi-class case, however, the transitions can occur among $2 \times K$ different neighbors. For instance, consider a setting in which we have two classes, $K = 2$, and the Markov chain is in state $(1, 1)$ (one active PEVs from each class). Then, there are four possible transitions from state (1,1), which include $(1, 0), (0, 1), (2, 1)$, and $(2, 2)$. Furthermore, the transition rates depend on the (λ_k, μ_k) pairs. To that end, the transition rates given in (6.28) represent PEV arrivals, whereas the transition rates in (6.29) account for new PEV arrivals. Similar to the single-class case, there cannot be any transitions beyond the neighboring states, which is reflected in (6.30).

6.4.1 Effective Demand

The analysis in Section 6.3.2 is valid for the multi-class case. However, computing the eigenvalues becomes infeasible when $K > 1$. Hence, we follow the decomposition approach proposed in [10] to efficiently evaluate the eigenvalues z_i and the coefficients α_i. The primary goal of this section is to compute ω_k, which is a deterministic quantity acting as a surrogate for the real aggregate stochastic demand [11].

We define parameter $\xi \triangleq 1 - \varepsilon^{1/B} \in [0, 1]$ to model the susceptibility of the storage unit to PEV demand patterns. This parameter is an essential part of computing ω_k. First, we consider large storage units and propose a scheme for computing ω_k, and then we will show that the proposed scheme is also applicable to any storage size. Assume that we are able to choose $\boldsymbol{N} = (N_1, \cdots, N_K)$ in a way that $\mathbb{P}(S_t > B) \leq \varepsilon$ for small ε and suppose that eigenvalues are relabeled so that $z_0 \geq z_1 \geq \ldots \geq 0$. Under these notations the following theorem holds.

Theorem 2. Let $\mathcal{B}(B, \varepsilon) = \{\boldsymbol{N} : \mathbb{P}(S_t \geq B) \leq \varepsilon\}$. For large storage size B and for small ε, we have

$$\lim_{\substack{\varepsilon \to 0 \\ B \to \infty}} \frac{\log \varepsilon}{B} \to \zeta \in [-\infty, 0] . \tag{6.33}$$

Furthermore, define

$$\tilde{\mathcal{B}} \triangleq \left\{ \boldsymbol{N} : \sum_k \omega_k(\xi) N_k < C \right\},$$

and

$$\bar{\mathcal{B}} \triangleq \left\{ \boldsymbol{N} : \sum_k \omega_k(\zeta) N_k \leq C \right\},$$

where the effective demand can be computed by [10]

$$\omega_k(\zeta) = \frac{\zeta R_k + \mu_k + \lambda_k - \sqrt{(\zeta R_k + \mu_k - \lambda_k)^2 + 4\lambda_k \mu_k}}{2\zeta}. \tag{6.34}$$

Then, $\tilde{\mathcal{B}} \subseteq \mathcal{B}(B, \varepsilon) \subseteq \bar{\mathcal{B}}$.

The interpretation of this theorem is that (6.4) holds if and only if

$$\sum_k \omega_k(\zeta) N_k \leq C\,.$$

6.5 Numerical Examples

6.5.1 Single-class Customers

In this section, we present various numerical examples to illustrate the system dynamics and explain how the introduced scheme can be used in designing charging facilities. We use the aforementioned normalized values (unit time is measured in μ^{-1} and unit demand is measured in peak demand R_p). We start by exploring the connections between the number of charger slots, ESS size (in $R_p\mu^{-1}$ units) and the corresponding underflow probability for a given system capacity C_t. Charge request rate per single charger λ is set to 2 (two PEVs arrive and request of size R_p arrives in unit time), and the average capacity above the mean demand per user is set to $\upsilon = 0.035$. Then the total system capacity becomes $C = 0.3683N$ units. In Figure 6.4, ESS sizing is evaluated for user population N ranging from 50 to

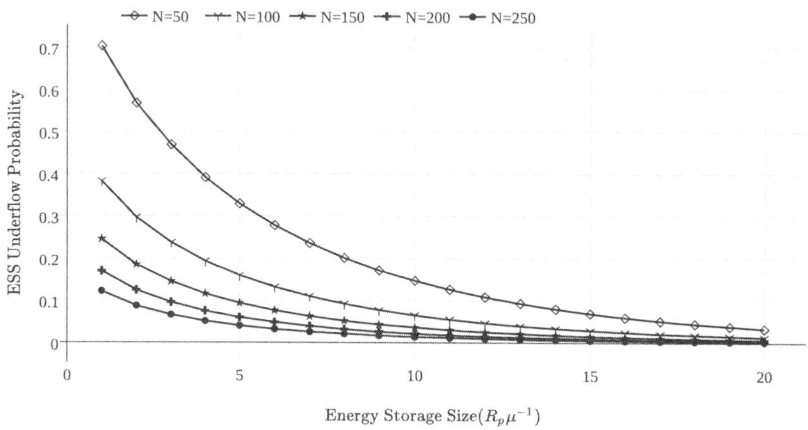

Figure 6.4 Energy storage sizing for parking lots with different sizes.

250. It is noteworthy that such numbers are very typical for parking lots located at convention centers, airports, and hospitals. The result presented in this figure can be interpreted and used for different purposes. First, given user population N, the system operator can choose the ESS size according to a certain underflow probability. For instance, for a large-scale PEV charging facility with $N = 200$ charging slots, accommodating 99% of the customer demands requires the ESS size be selected as $B = 15 \times R_p \times \mu^{-1} = 15 \times 6.6 \times 0.5 = 49.5$ kWh (assuming 6.6-kW level 2 chargers). One important observation is that as the user population increases the required ESS size reduces due to the increase in the multiplexing gains. Another important observation is that instead of sizing the ESS to meet the entire customer demand, by rejecting only a small fraction of the customers, considerable savings in the storage size, and subsequently aggregate system costs, can be achieved. It is noteworthy that the choice of energy storage is usually application-dependent. For instance, when the charging facility is located in a large metropolitan downtown area, the size of the storage can become problematic as the growth of the physical size might be prohibitive. In order to demonstrate this point, assume that the station operator aims to install a storage unit of size $B = 64$ kWh and chooses the Maxwell BMOD0083P048 module (each has 26.6-Wh capacity). This requires 2406 of such modules with the aggregate volume 24 m^3. This example illustrates how physical

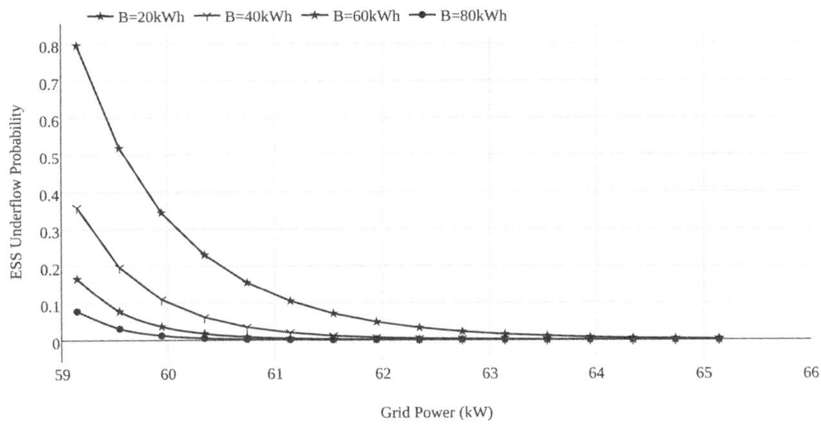

Figure 6.5 Energy storage underflow probability for varying grid power and storage size.

space can be problematic in real-world applications. One potential solution is to place these modules underground. Space issues become less severe when the storage units can be accommodated in locations where space limitations are less of a concern, such as inside large shopping malls or along highways.

Another design consideration pertains to the circumstances in which the charging station already exists, meaning that N is given. In such circumstances, the design goal boils down to computing the optimum grid power and battery size combination. Figure 6.5 evaluates such a scenario with $N = 200$ and level 2 charging slots. In a real-world scenario, there exist constraints on the available power that can be drawn from the grid, due to, for example transformer rating, in which case the design goal reduces to calculating the storage size for a target performance metric.

Recall that the ultimate objective is to minimize the ESS size so that certain performance measures are guaranteed. In Figure 6.6, we compute the percentage of savings in ESS size with respect to peak demand allocation, which is $N \times R_p$. The results demonstrate that as the size of the charging station grows, the saving in storage size increases due to resource-pooling.

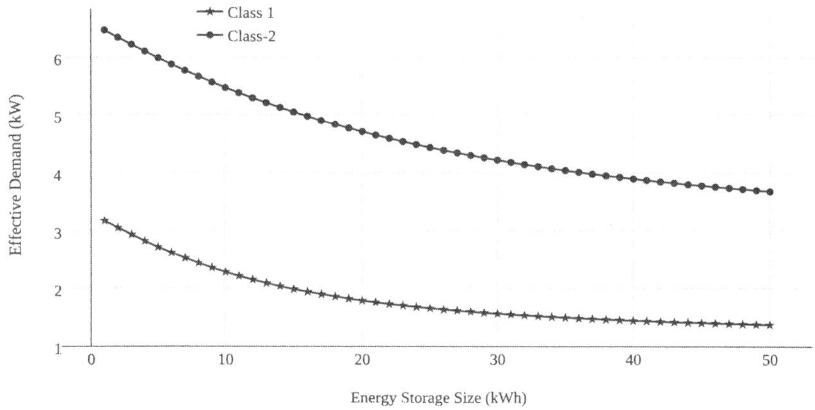

Figure 6.6 Percentage of savings in ESS size for varying charging station sizes.

6.5.2 Multi-class Customers

Next, we consider a case in which customers can request different demand levels, or they are equipped with different charging technologies, and as a result constitute different customer classes. Recall that the objective in such settings is to compute the ω_k (i.e., effective demand) parameter for each class so that ω_k replaces R_k and the overflow probability targets are satisfied. For example, consider a scenario in which the PEV charger can generate four different levels of demand, with the following parameters $\boldsymbol{\lambda} = \{0.3, 0.5, 0.7, 0.9\}$, $\boldsymbol{\mu} = \{1, 1, 1, 1\}$, and $\boldsymbol{R}^k = \{3, 4, 5, 6\}$ (in kW). Then for the target of $\varepsilon = 10^{-4}$, and ESS size of $B = 20$ kWh, the corresponding effective demands are $\boldsymbol{\omega} = \{1.28, 2.24, 3.22, 4.21\}$. Alternatively, the network operator can find the minimum storage size with respect to available grid resources and target performance guarantees. For convenience in notations, the remaining numerical results consider two classes of customers, and to mimic the level 1 and level 2 chargers we select the following parameters $\lambda_1 = 0.5$, $\mu_1 = 1$, $R_1 = 3.3$ kW and $\lambda_2 = 0.7$, $\mu_2 = 1$, $R_2 = 6.6$ kW. Our first evaluation shows the acceptance set for $C = 100$ kW and $B = 25$ kWh, for a target overflow probability of $\varepsilon = 0.001$. Results presented in Figure 6.7 show that the operator can determine the number of customers from the shaded section. Clearly, since the

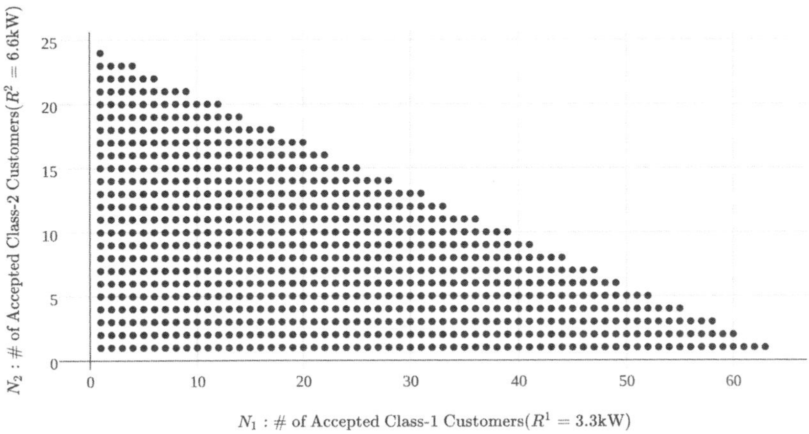

Figure 6.7 Acceptance region for K=2.

demand for Class 1 customers is less than that required by the customers in the other class, the system operator can accept more customers from Class 1. Such evaluations are critical, especially in the early stages of system planning.

In another setting, we compute the effective demand for varying storage sizes. In this setting the target underflow probability is set as $\varepsilon = 5 \times 10^{-4}$ and customer populations are $N_1 = 50$ and $N_2 = 50$. The results depicted in Figure 6.8 show that instead of provisioning the system according to peak demand ($R_1 = 3.3$ kW and $R_2 = 6.6$ kW), adopting effective demands lowers the provisioning of resources substantially. Our last results are on the ESS sizing for multiple classes. For the same set of parameters, we evaluate the storage size with respect to the power drawn from the grid. Similarly to the single-class setting, these results (shown in Figure 6.9 can be used to size the storage units for a given grid power or for a given storage size.

6.6 Chapter Summary

In this chapter we formalized a framework for designing large-scale charging stations, where the key observations are:

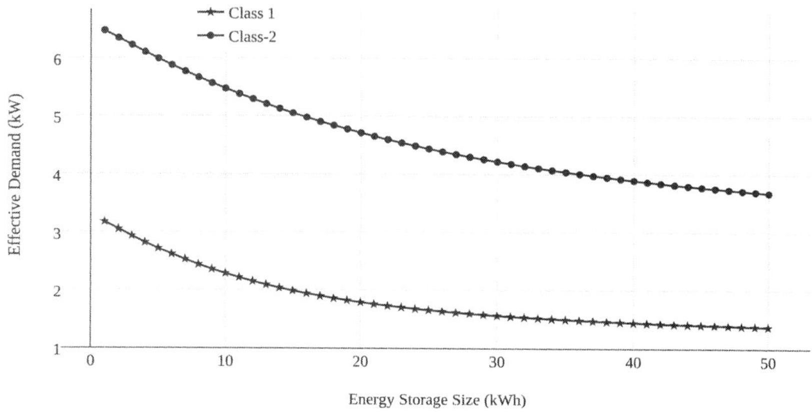

Figure 6.8 Effective demand calculation for K=2.

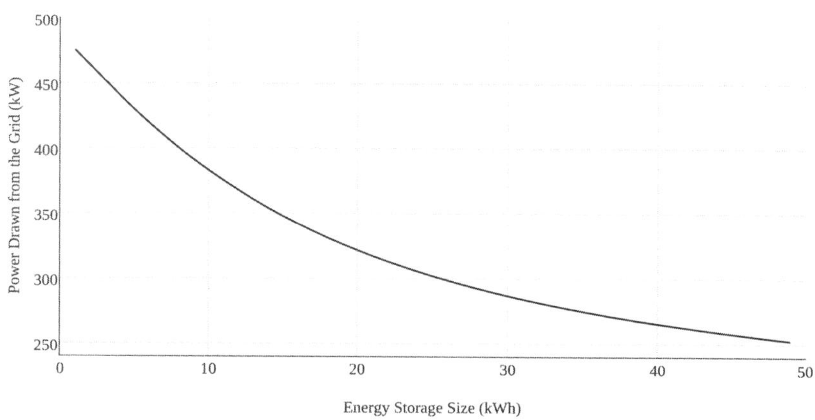

Figure 6.9 ESS sizing for K=2.

1. Charging facilities equipped with energy storage units can lower the stress on the grid, which is critical during peak hours. The proposed station architecture enables station operators to provision the required resources (grid power and storage size) according to the stochastic PEV demands.

2. The results show that sharing-based storage solutions are economically viable and foster the deployment of public charging stations. The choice of the storage size is often application-dependent and the physical size of the storage unit can also play a significant role in certain congested locations.

References

[1] Bayram, I.S., et. al.,"Electric Power Resource Provisioning for Large Scale Public EV Charging Facilities", *Proc. IEEE International Conference on Smart Grid Communications*, Vancouver, CA, Oct., 2013, pp. 133-138.

[2] Nicholas, M., Tal, G. and Woodjack, J. "California Statewide Charging Survey: What Drivers Want?," *92nd Annual Meeting of the Transportation Research Board*, Washington DC, Jan. 2013.

[3] Wang, Gui, et al., "Dynamic Competitive Equilibria in Electricity Markets," *Control and Optimization Methods for Electric Smart Grids*, pp. 3-33, A. Chakrabortty and M. Ilic (Eds), New York: Springer-Verlag, 2012.

[4] Bayram, I.S., et al., "A Stochastic Sizing Approach for Sharing-Based Energy Storage Applications", *IEEE Transactions on Smart Grid*, Vol. 8, No. 3, pp.1075-1084.

[5] Bayram, I.S., et al., "Energy Storage Sizing for Peak Hour Utility Applications," *Proc. IEEE International Conference on Communications* , London, Jun. 2015, pp. 770-775.

[6] Ardakanian, O., Keshav, S., and Rosenberg, C., "Markovian Models for Home Electricity Consumption," *Proc. ACM SIGCOMM Workshop on Green Networking*, Toronto, CA, Aug., 2011, pp. 31-36.

[7] Anick, David, et al., "Stochastic Theory of a Data-Handling System with Multiple Sources," *Bell Systems Technical Journal*, Vol. 61, No. 8, 1982, pp. 1871-1894.

[8] Schwartz, M., *Broadband Integrated Networks*, Upple Saddle River, NJ: Prentice Hall PTR, 1996.

[9] Morrison, J. A., "Asymptotic Analysis of a Data-handling System with Many Sources," *SIAM Journal on Applied Mathematics*, Vol. 49, No. 2,1989, pp. 617-637.

[10] Gibbens, J. and Hunt, PJ, "Effective Bandwidths for the Multi-type UAS Channel", *Queueing Systems*, Vol. 9, No. 2, 1991.

[11] Kelly, F. "Effective Bandwidths at Multi-class Queues," *Queueing Systems*, Vol. 9, No. 2, 1991.

Chapter 7

Capacity Planning for Small-Scale PEV Charging Stations

7.1 Introduction

In the previous chapter, we introduced a capacity planning framework for large-scale charging stations equipped with level 1 and level 2 chargers, which suffice to charge a typical PEV battery within several hours. Such facilities are envisioned to provide service to customers who can remain stationary and possibly follow routine commute patterns (e.g., daily commute to workplace). In order to compete with gas-powered vehicles, deploying charging facilities that expand access to charging stations beyond level 2 charging can fully play a critical role. Apparently, the critical design criteria in such systems include (1) easy public access in highly populated urban regions as well as main roads and highway exits, and (2) short service duration empowered by high charging power. The first set of criteria pertains to station placement problem, which requires investigating transportation and power networks in parallel. Such designs will be discussed in the upcoming chapters. The second design element can be addressed by using DC fast charging technologies. As discussed in Chapter 3, fast chargers can provide a power transfer rate in the range of $35-50$ kW, and hence, depending on the battery packs, the charging duration varies in the range of $15-30$ minutes.

Increased availability of public fast-charging stations enhances customers' confidence and accelerates PEV adoption. More specifically, fast-charging stations will serve drivers who do not have access to garage charging and, therefore, they will act as complementary services to extend the all-electric range. For instance, half of the PEV owners in southern California require recharging during their daily commutes. On the other hand, the current limitations of the power grids suggest that charging stations could only be implemented at a small scale and serve $10-12$ customers simultaneously [6]. In particular, as of 2016 the largest capacity of an operating fast-charging station is 350 kW, which can also be translated into a station of $7-10$ physical chargers depending on the regional standards and regulations. Moreover, recent advances in energy storage technology enable system operators to employ large energy storage units to expand the service capacity. To that end, in this chapter we investigate the capacity planning aspects of a public fast-charging station. According to the U.S. Department of Energy Alternative Fuel Data Center, more than 1600 fast chargers are deployed in the United States. European countries are also showing a great deal of interest in expanding the size of fast-charging networks. In Netherlands, for instance, there are 50 fast-charging locations and this number is expected to reach 130 by the end of 2017, which will enable drivers to travel across the country with pure electric miles [2].

It is noteworthy that even though fast-charging networks have critical roles, there exists the concern that PEV demand can potentially exert additional strains on the grid due to the large load they impose. The extent of their impacts depends on the regional density of PEV users, charging requirements, and time of service, and could lead to grid instabilities. The local system imbalances can trigger cascading failures at a larger scale which can lead to regional blackouts. Therefore, planning and operating such fast-charging infrastructures should be investigated rigorously. In this chapter, we propose a PEV fast-charging station architecture equipped with a local energy storage system to meet the customer demands. Moreover, we introduce a stochastic model to capture the station's operational characteristics, which naturally couples the capacity planning problem with the target rates at which the customers are served. We introduce the notions of outage probability' and blocking probability as stochastic measures for evaluating the performance of the charging stations. The main goal is to ensure grid reliability via flattening demand fluctuations with storage units.

Studies on charging station architectures are quite limited, especially when taking into account the significance of this technology [3–6]. Some of the major studied include the work in [3], which proposes a station design with a DC bus system. The charging station is equipped with an energy storage unit to lower the strain on the distribution grid. Moreover, the ESS sizing problem was solved by Monte Carlo simulations according to average customer load. The study in [4] focuses on the power electronics component and proposes an architecture that can fill up a typical PEV battery within $10 - 15$ minutes. The studies presented in [7] and [8] propose charging station architectures that involve two storage units, including a flywheel unit and a super-capacitor unit, and have the ultimate goal of minimizing the charging duration.

7.2 System Description

Due to the typical size and capacity of fast-charging facilities, it is assumed that the utilities consider such facilities as commercial and industrial customers. The critical design principle in such stations, therefore, is to minimize the demand variations as much as possible to ensure power grid reliability. To address this issue, customer demands are served by the power drawn from the grid as well as a local energy storage system with the following operating structures:

- Since fast-charging stations represent commercial-size loads, it is assumed that station operators draw long-term contracts with the utilities, where predetermined power levels are agreed upon in favor of lower tariffs. Therefore, stations draw constant power from the grid.
- An energy storage system is employed in the stations to smooth stochastic customer demands. During peak or rush hours, energy storage in the ESS can be used to charge more vehicles. On the other hand, during off-peak hours the storage unit can be charged with the grid power. Even though the primary goal of the storage unit is to serve customers, it can also be used to participate in energy markets to sell electricity to make extra profit.[1]
- The service times depend on the charging power rate, power electronics, and battery technology. Nevertheless, the average charging takes around 30 minutes for mainstream PEV models. Therefore, it is reasonable to assume that arriving customers prefer not to wait in the system (or there could be a few slots for waiting and this case will be investigated in the next chapters); therefore, it is reasonable to model the stations as bufferless systems. In such systems, outage or blocking probability arise as natural performance metrics.

1 However, for this basic model we do not consider this case.

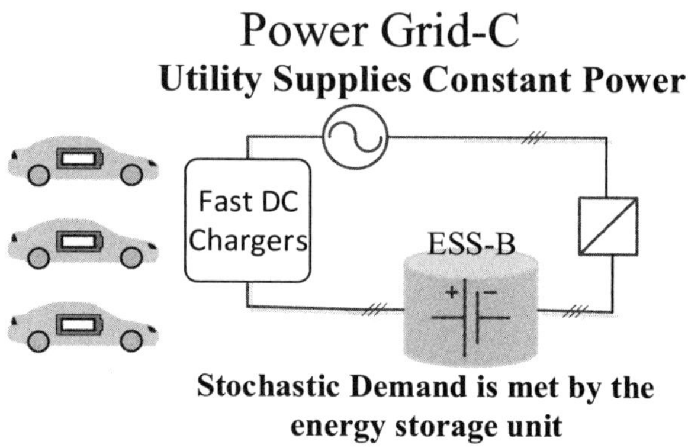

Figure 7.1 Small-scale charging station model.

A schematic overview of the charging station architecture is depicted in Figure 7.1.

7.2.1 Stochastic Model

Based on the specification described above, the proposed station architecture operates under the following setting. The constant power drawn from the grid is discretized to C slots, meaning that C vehicles can be concurrently accommodated by the grid power. In a similar way, it is assumed that the energy storage can charge up to B vehicles when it is in a fully charged state. When an arriving customer finds all the station resources (grid and storage) being committed to others, it will be blocked and leaves the system. This strategy is adopted to avoid peaks in power demands.

In the proposed framework, PEV arrivals are modeled by a Poisson process with parameter λ. This assumption is widely used in the literature (see [5, 6, 9] and references therein) because in the Poisson process customers' arrivals behave statistically independently of each other.[2] Moreover, as discussed in Chapter 2, there are a variety of different PEV models with different battery packs and charging durations. Therefore, the service time of customers is assumed to be exponentially distributed with rate μ. Similarly, the charging duration of one slot in the energy storage device is also exponentially distributed with parameter ν. It is noteworthy that this rate depends on the employed storage technology parameters such as

2 Refer to appendix for justification of this assumption.

power rating and charging efficiency. In this framework, the arriving customers are first served by the grid power, and when the number of customers seeking service exceeds C, the local ESS also starts providing the additional needed demand. It is noteworthy that the station cannot serve more than $C+B$ customers simultaneously. Therefore, the goal is to provide the customers a reasonable quality of service, which varies with the grid power and the storage size.

Based on the aforementioned setting and assumptions, the single charging station can be modeled by a continuous-time Markov chain with 2-dimensional state space. In Figure 7.2 the state space, along with the transition rates are depicted. The horizontal transitions correspond to customer arrivals and departures, while the vertical ones model the local ESS charge and discharge events. For instance, the transition rates shown with bold font represent the case where the customer is served by the ESS, and the transition rates shown by ν depict the local ESS charging events.

The total number of states is in the Markov chain in Figure 7.2 is

$$S = (C+1)(B+1) + \sum_{i=1}^{B} i \, .$$

It is easy to see that a unique steady state distribution $\boldsymbol{\pi} = (\pi_1, \pi_2, \ldots, \pi_S)$ exists, which can be calculated by solving

$$\boldsymbol{\pi}\boldsymbol{Q} = \boldsymbol{0} \quad \text{and} \quad \sum_{i=1}^{S} \pi_i = 1 \, . \tag{7.1}$$

In the above equation, the $S \times S$ matrix $\boldsymbol{Q}$ contains the transition rates and is called the infinitesimal generator matrix. Note that the elements of $\boldsymbol{Q}$ satisfy

$$q_{ij} \geq 0 \, , \quad \forall i \neq j \quad \text{and} \quad q_{ii} = -\sum_{j \neq i} q_{ij} \, , \quad \forall i \, . \tag{7.2}$$

It is important to note that the rightmost states, namely

$$(C, 0), (C+1, 1), \ldots, (C+B, B) \, ,$$

correspond to blocking states. Hence, when we sort the states from 1 to S starting from $(0,0)$ to $(C+B, B)$; that is, 1 is assigned to $(0,0)$, and $(C+2)$ is assigned

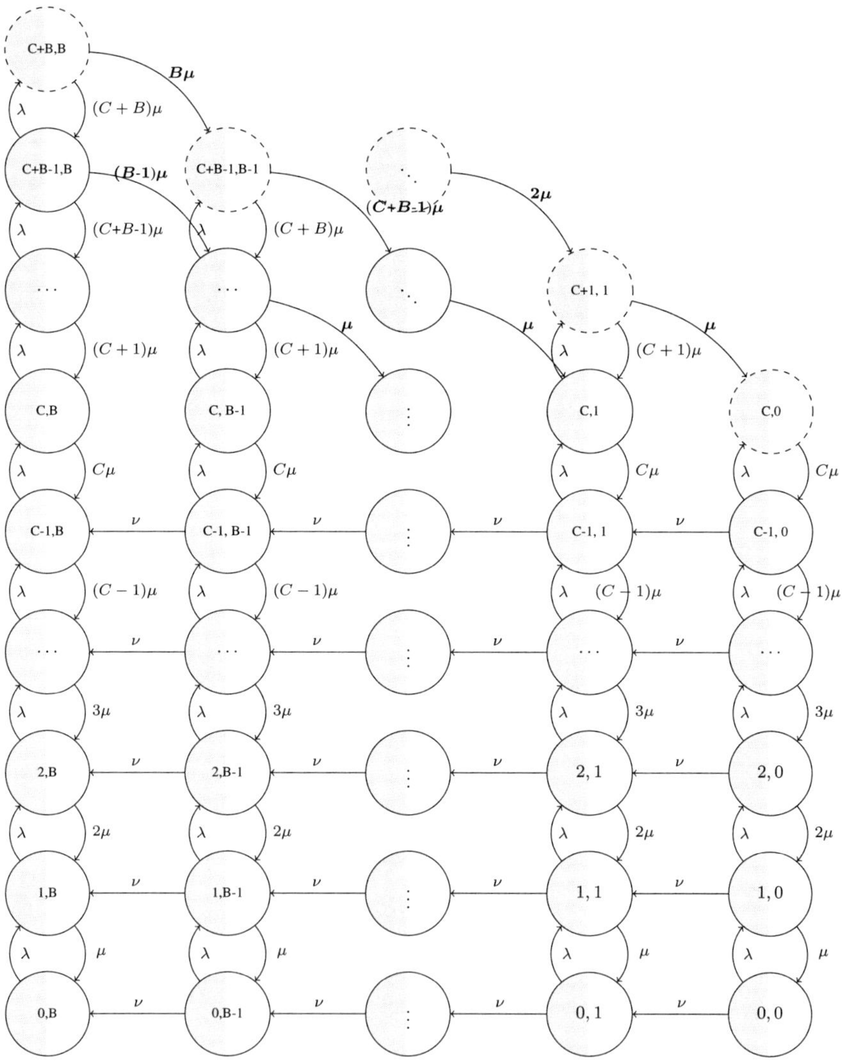

Figure 7.2 Markov chain model.

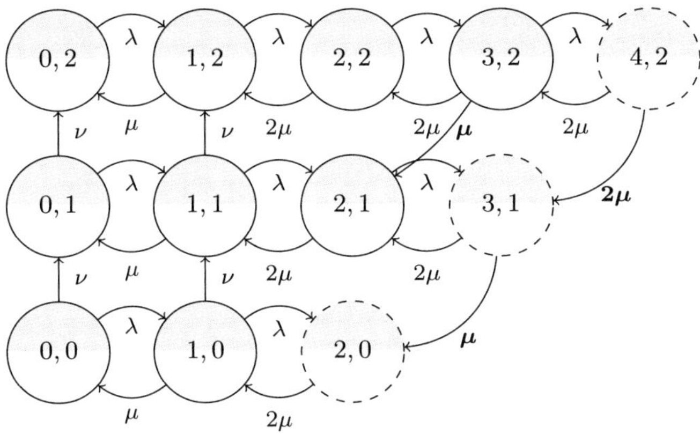

Figure 7.3 Markov chain model for toy example.

to $(0, 1)$, the blocking probability can be calculated according to

$$\sum_{i=1}^{C} \pi\left(\frac{i(i+2C+1)}{2}\right).$$

Moreover, the infinitesimal generator matrix can be constructed as

$$\boldsymbol{Q} = \begin{pmatrix} -(\lambda+\nu) & \lambda & \cdots & 0 \\ \mu & -(\lambda+\nu+\mu) & \cdots & 0 \\ \vdots & \vdots & \ddots & \vdots \\ 0 & 0 & \cdots & -(C+B)\mu \end{pmatrix}. \tag{7.3}$$

7.2.2 Toy Example

We provide a toy example to better explain the Markov chain model. Let us assume that the grid power and the energy storage unit can charge two customers separately (i.e., $C = B = 2$). Hence, the total number of states in the Markov chain becomes $C = 12$, and the corresponding Markov chain is illustrated in Figure 7.3.

$$
\boldsymbol{Q} = \begin{pmatrix}
-\lambda-\nu & \lambda & 0 & \nu & 0 & 0 & 0 & 0 & 0 & 0 & 0 \\
\mu & -\mu-\lambda-\nu & \lambda & 0 & \nu & 0 & 0 & 0 & 0 & 0 & 0 \\
0 & 2\mu & -2\mu & 0 & 0 & 0 & 0 & 0 & 0 & 0 & 0 \\
0 & 0 & 0 & -\nu-\lambda & \lambda & 0 & 0 & \nu & 0 & 0 & 0 \\
\ddots & \ddots & & \cdots & & & & & & \ddots & \ddots \\
0 & 0 & 0 & 0 & 0 & 0 & 2\mu & 0 & 0 & 2\mu & -4\mu
\end{pmatrix}
\tag{7.4}
$$

Similarly to the generic Markov chain, the dotted states correspond to blocking states and the bold transitions show the case when the customer demands are served by the local ESS. Also, as given in (7.1) the steady-state probabilities can be calculated by solving a system of linear equations. For instance, let us assume that on average two customers arrive at each hour, hence $\lambda = 2$. Moreover, on average it takes 30 minutes to serve one vehicle and the same rate is assumed to charge a slot of a vehicle in the local ESS, then $\nu = \mu = 2$. By solving the system of linear equations, the corresponding steady state probabilities become $\boldsymbol{\pi} = (0.0062,\ 0.0125,\ 0.0156,\ 0.0211,\ 0.0361,\ 0.0373, 0.0188,\ 0.0345,\ 0.032473,\ 0.0324, 0.1337,\ 0.0382,\ 0.0095)$. Note that the blocking probability is a function of system parameters and is calculated by summing up the probability of blocking states; that is $f(C, B, \lambda, \mu, \nu) = 0.0156 + 0.0188 + 0.0095 = 0.0439$. This means that under this station setting, on average 4.39% of the customers will not receive service. In order to enhance the service quality (i.e., reduce the outage rate), station operators must either increase the storage size or the power provided by the grid.

7.3 Profit Model

Performance assessments provide insights into the gains captured by the proposed performance metric. In order to incentivize the station operator to provide an acceptable QoS to the customers (e.g., typically less than 5% outage probability), we formalize a profit model that relates the stochastic model to cost parameters and guides station operator to provision station resources optimally.

The principles of the profit model are as follows. Each customer who is accepted for service pays a constant fee. As discussed in Chapter 5, there exist a variety of pricing schemes for this purpose. For instance, in a flat pricing scheme, the station operator charges a fee to cover its operating cost and the prices are kept constant at all times. However, such pricing schemes fail to alleviate congestion. On the other hand, there are the time-of-use pricing schemes widely employed by utilities, in which different rates are applied at different time durations. In our model, the station operator obtains R_g amount of revenue at *grid charging states* denoted by $\rho^{(g)} = \{(i,j) : 0 \leq i \leq C, 0 \leq j \leq B\}$, while the service fee is R_b during the *energy storage charging states*, denoted by $\rho^{(b)} = \{(i,j) : C + j \leq i \leq C + B, 1 \leq j \leq B\}$. If a customer cannot get service (i.e., it is blocked), the station operator pays a penalty to each customer, which is denoted by R_p. Notice that the blocking events occur when the station is in a blocking state (denoted by $\rho^{(bl)}$). The reasons underlying the penalty fee include (1) blocking events leads to dissatisfied customers and degrades the reputation of the service provider, (2) it enables a control mechanism to guarantee achieving QoS targets, and (3) it enables resource provisioning for the station operator. Moreover, the relationship between the rates is $R_p > R_b \geq R_g$. Note that $R_b > R_g$ reflects TOU pricing, since higher prices are charged during peak periods, while flat pricing is applied when $R_b = R_g$. Furthermore, let C_0 denote the fixed installation cost and BC_a denote the acquisition cost proportional to the size of the storage unit. Based on these definitions, the station profit model can be written as

$$\begin{aligned} P &= \sum_{s \in \rho^{(g)}} R_g i(s)\pi(s) + \sum_{s \in \rho^{(b)}} R_b i(s)\pi(s) \\ &- C_0 - BC_a - \sum_{s \in \rho^{(bl)}} R_p i(s)\pi(s) , \end{aligned} \tag{7.5}$$

where $i(s)$ denotes the number of PEVs at state s.

7.4 Performance Evaluation

We evaluate the charging station performance (fraction of the PEVs it can charge), under varying system parameters. First, we remark that the average vehicle charging rate (μ) and average ESS charging rate (ν) are functions of power rating and efficiency of the storage unit. Therefore, these two parameters are considered as technological constraints and are assumed to be constant through the evaluations. As given in the toy example, μ and ν are set to 2, meaning that on average one charger can serve two vehicles in an hour.

The system performance is evaluated for four different grid power settings; namely $C = 4, 6, 8$ and 10 for arrival rate in the range $\lambda \in [0.1, 15]$ with increment intervals of 0.1, and energy storage unit in the range $B \in [1, 10]$. The results depicted in Figures 7.4 and 7.5 provide practical information about the system dynamics. For a given traffic regime and a target blocking performance, the station operator can choose the right values for grid power and the storage size. It is easy to observe that the grid power lowers the chance of blocking events. However, the impact of local energy storage unit is not the same. For instance, for $C = 4$, under the high traffic regime (e.g., $\lambda \geq 10$), the grid power is mostly used to serve vehicles, hence, the local ESS remains discharged.

Our final evaluation pertains to the station profit model, which was described in the previous section. In our setting, μ and ν remain the same and set to 2. Since our primary goal in this chapter is to provide a basic modeling approach, it is assumed that the station employs flat pricing, meaning that $R_g = R_b = 1$. If a customer is blocked, the station pays a penalty (e.g., a commitment to a future service at a reduced rate) of $R_p = 1.1$. The cost parameters C_0 and C_a are normalized and set to 2.5 and 0.001, respectively. We perform the calculations for two different station settings $C = 6, B = 5$ and $C = 6, B = 6$ and the results are shown in Figure 7.6.

The results convey practical information for the station operation. First, the station profit is negative under the light traffic regimes ($\lambda \leq 5$) because the cost of acquiring local energy storage substantially exceeds the revenue generated. Second, the station profit decreases because the station starts to provide degraded QoS and the penalty cost outweighs the revenue obtained by the charging service. Therefore, both case studies indicate that in order to maximize the profits, station capacity should be provisioned according to the customer demand.

Furthermore, it is noteworthy that the station model is designed to meet the peak-hour customer demand. For instance, for the peak hour demand is $\lambda = 8.5$, then the station operator can choose C and B based on the QoS target and the profit model. For the given case, the choices $C = 6$ and $B = 6$ satisfy the QoS target of 1.5%. A natural question to ask is what happens if the arrival rates are non-homogenous, meaning that λ varies over time. In this case, the local ESS could be used to provide ancillary services in the energy markets. Another option is to provide average QoS guarantees for the longer term. This means that the QoS may be violated at certain time slots, but a target QoS for long-term average can be sustained.

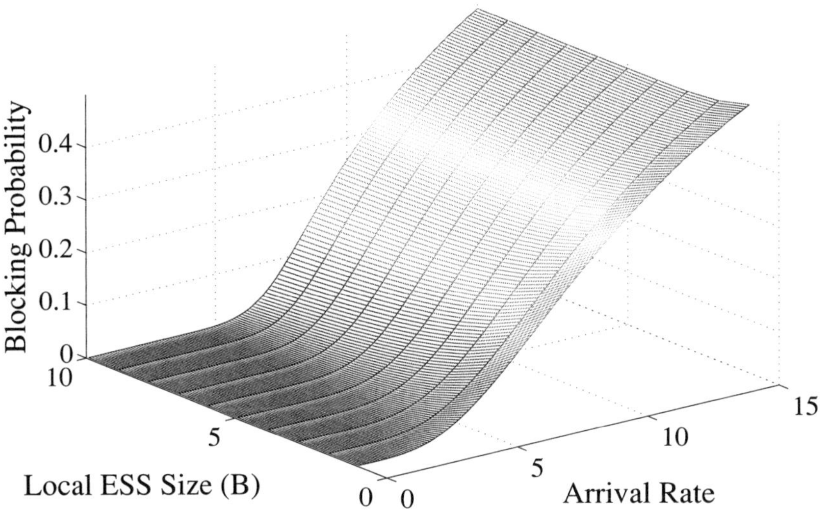

(a) Performance evaluation for C=4.

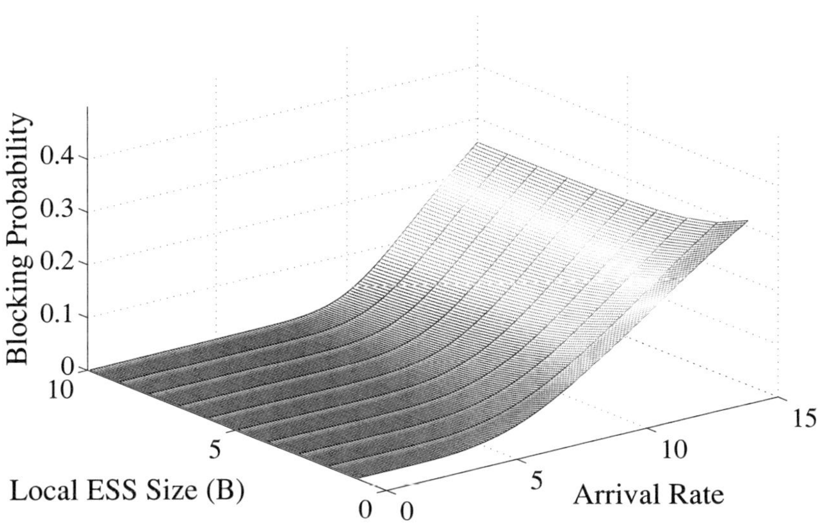

(b) Performance evaluation for C=6.

Figure 7.4 Small-scale charging station performance evaluation under varying grid power: example 1.

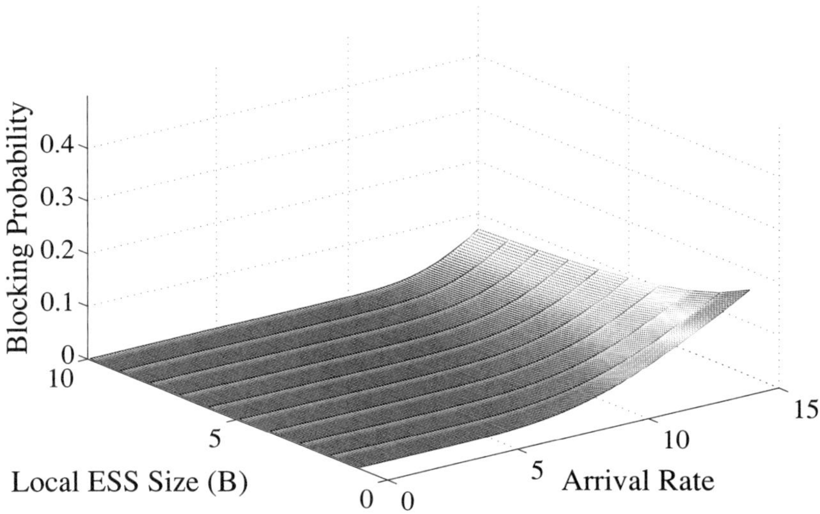

(a) Performance evaluation for C=8.

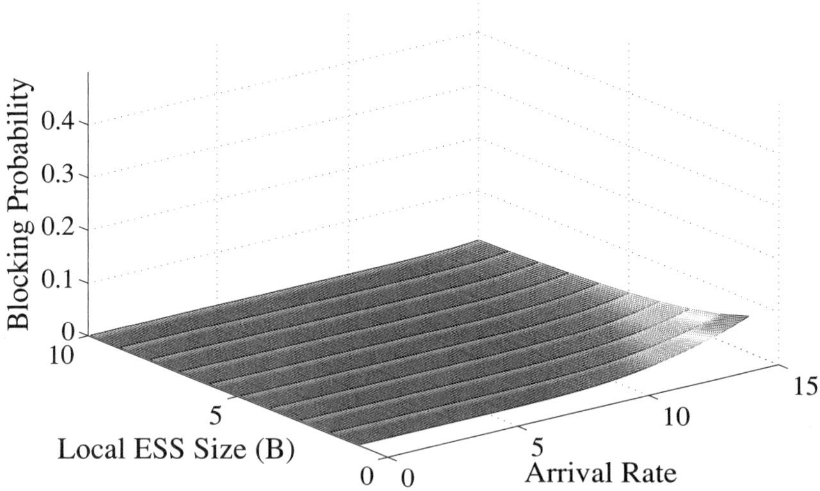

(b) Performance evaluation for C=10.

Figure 7.5 Small scale charging station performance evaluation under varying grid power: example 2.

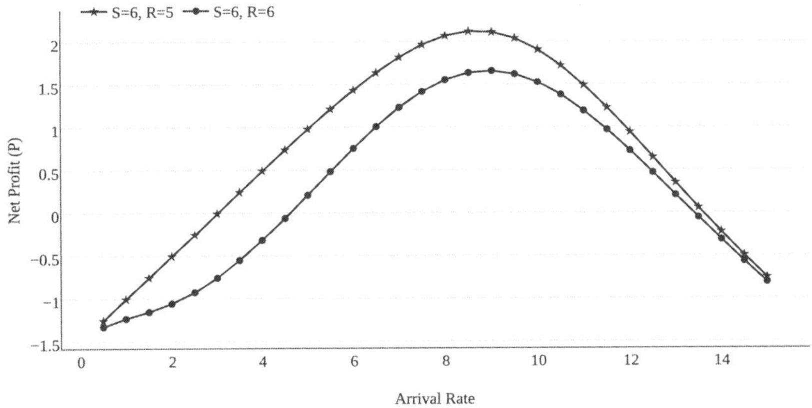

Figure 7.6 Charging station profit model.

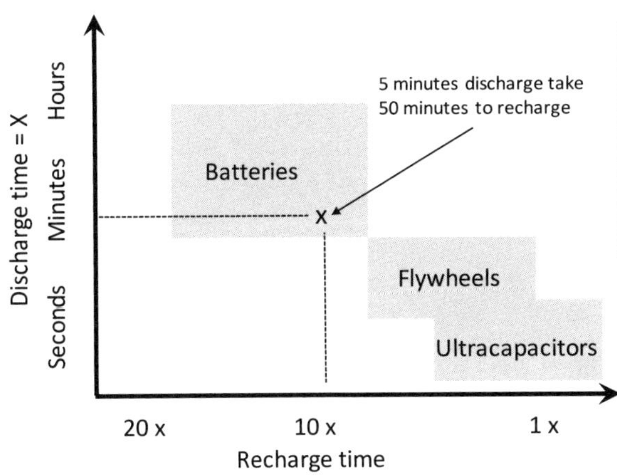

Technology	Efficiency
Battery	60%-85%
Flywheel	93%
Ultracapacitors	95%
Hydrogen Fuel Cell	59%

Figure 7.7 Candidate local energy storage systems landscape.

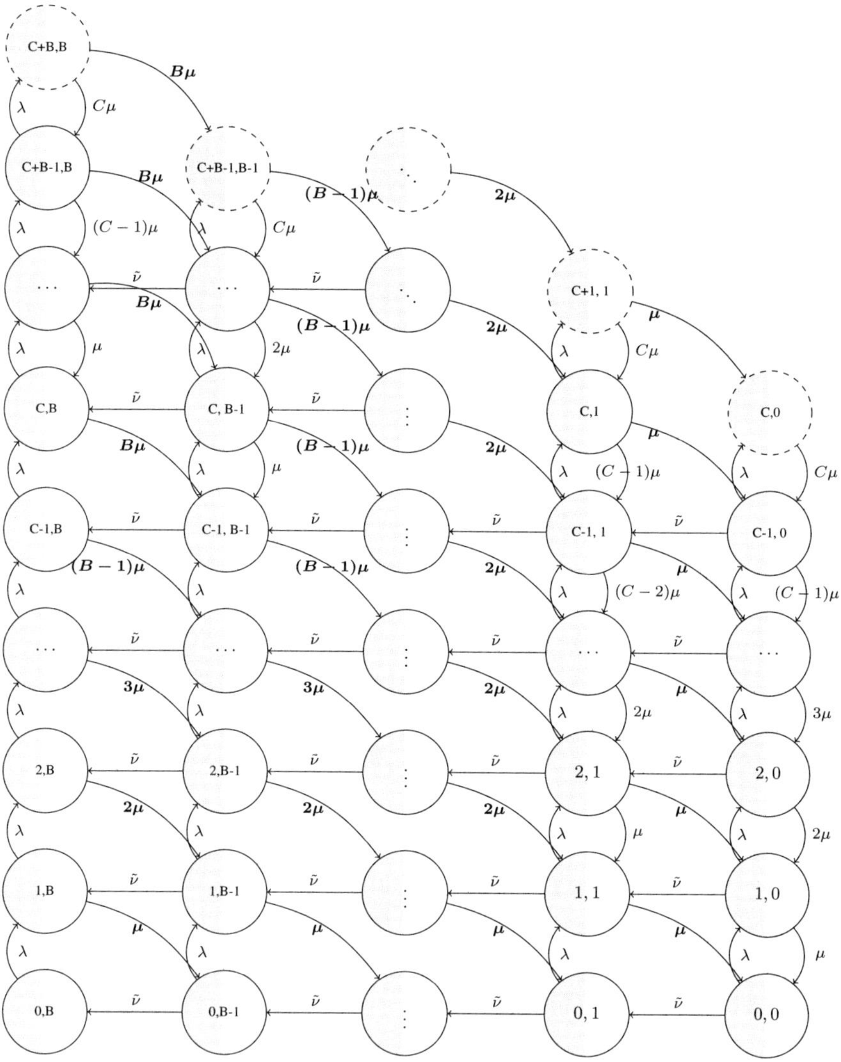

Figure 7.8 Markov chain model for the charging from the ESS first ($C = B$).

7.5 Stations with Competing Energy Storage Technologies

In the previous sections, a generic charging station architecture equipped with a local energy storage device was presented together with stochastic operation principles. The latter incorporated results from queuing theory and finance for the station's operator, thus allowing one to evaluate the system performance with respect to different performance metrics, including blocking probabilities for customers receiving service and profit realized by the operator.

The operation principle of the model gives priority to accommodating vehicles with the grid power, while the primary goal of the local energy storage unit is to serve customers during peak periods and alleviate the variability in customer demands. The proposed basic model assumes that the local energy storage technology also uses electrochemical storage units and the numerical evaluations consider the setting $\mu = \nu$. However, as discussed in the introduction, different local energy storage technologies can be employed at the charging facility. The critical aspect of such storage units is that their charging and discharging rates and their efficiency levels can be higher than those of onboard PEV batteries. Therefore, the local ESS charging rate ν could be higher than the charging rate of a vehicle denoted by μ. Recall that station performance is measured by the blocking probabilities, which is the percentage of time the system stays in the blocking states. This also means that more energy can be stored in the local storage unit and there will be more vertical transitions in the Markov chain, which facilitates reducing the system blocking probability. The following example signifies the effects of different charging technologies. Consider a fast-charging station that can charge a PEV, whose battery has efficiency $\eta = 0.9$, in 30 minutes using maximum power rating, $C_{\text{PR}} = 1$ with the charging rate parameter $\mu = 2$. Next, consider the same system, which is also equipped with a storage unit with $\nu = 4$, the same efficiency, but with a higher power rating $C_{\text{PR}} = 2$ than the PEV's Li-ion battery. Note that during the same amount of time, the storage unit can store the energy needed to serve two PEVs. Hence, the new charging rate $\tilde{\nu}$ becomes a function of the power rating and the efficiency of the local storage unit that is given by $\tilde{\nu} = g(\hat{C}, \eta)$, with $\hat{C} \leq C_{\text{PR}}$ being the available power. An overview of candidate energy storage system characteristics is shown in Figure 7.7.

In this section, we extend the model presented in the previous section by incorporating the dynamics of the candidate energy storage systems. The constant ESS charging rate ν is replaced with a state-dependent charging rate $\tilde{\nu}$. Moreover,

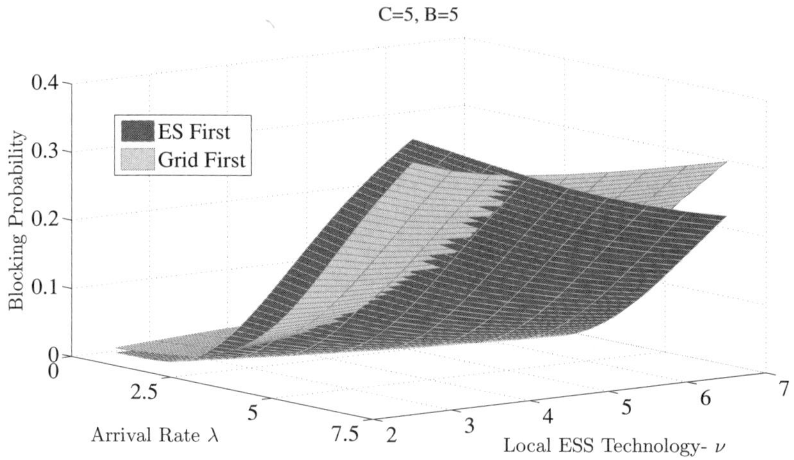

Figure 7.9 Comparison of two charging strategies.

we employ a new strategy in which the PEVs are first charged by the local energy storage. If at some point, available energy cannot meet customer demands, the power drawn from the grid will be used. The new strategy is called *charge from the ESS first*, while the previous one icalled *charge from the grid first*. Other assumptions on the transition rates for the Markovian model remain the same. To that end, the 2-dimensional Markov chain for the new strategy is presented in Figure 7.8.

It is noteworthy that the steady state probabilities can be calculated in a similar fashion, but the infinitesimal generator matrix needs to be updated. The new model contains new transitions between the vertical states, which are shown in bold font. For instance, consider the previously discussed toy example, where $C = B = 2$ with 12 states. Then, the fifth row of the new $\boldsymbol{Q}$ matrix, corresponding to state $(1, 1)$, is $(\mu, 0, 0, 0, -(\mu + \lambda + \nu), \lambda, 0, 0, \nu, 0, 0, 0)$, and other rows could be constructed in a similar way.

The two strategies are evaluated for varying arrival rate (λ) and energy storage technologies ($\hat{\nu}$). The results depicted in Figure 7.9 provide interesting insights. If the station operator employs a fast storage unit, then the strategy that first charges from the local ESS outperform the strategy that first charges from the grid first. For

Table 7.1
RSM Input Parameters

Parameter	Interval	Increments	Type
S	[1,15]	1	Integer
R	[1,15]	1	Integer
λ	[0.25,30]	0.25	Float
ν	[2,10]	1	Integer

slow ESS technologies $\nu \leq 3$, the first strategy performs better, and thereby serves more customers. Hence, depending on the local ESS technology, the station operator can choose an optimal operating strategy to serve a larger number of customers.

7.6 System Metamodeling Approach

In the previous sections, we provided numerical methods to calculate the PEV blocking probabilities. In this section, we provide analytical expressions for the optimization problems involved in the proposed charging station model. To this end, we employ the response surface methodology (RSM) to determine the blocking probability function P_B in terms of C, B, ν, and λ [10].[3]. For the RSM the input parameters are presented in Table 7.1 and μ is set to 2, which is the parameter for the Li-ion batteries commonly used by PEVs.

The key impediment facing this approach is that the blocking probabilities have to reside in the $[0, 1]$ interval, whereas the RSM model predictions do not enforce any limitations on these measures. Therefore, we fit the RSM model to the logit transformation ($y = \log(x/(1 - x))$) of P_B and then use the inverse logit ($x = 1/(1 + e^{-y})$) transformation to obtain the final results. The regression model, in which the response variable corresponds to logit is

3 The RSM is computed for the "charge from the grid first" strategy, but it can also be applied to the other strategy as well

$$
\begin{aligned}
P_B(C,B,\lambda,\nu) &= -3.990 - 2.666C - 1.6152B - 0.1492\nu \\
&+3.840\lambda - 0.0645CB - 0.002C\nu + 0.209C\lambda - 0.0078B\nu \\
&+0.094B\lambda + 0.003\nu\lambda - 0.0175C^2 + 0.055B^2 + 0.0089\nu^2 \\
&-0.271\lambda^2 . \qquad (7.6)
\end{aligned}
$$

Obviously, the blocking probability occurs only when the arrival rate is strictly positive.

For the above regression model, the R-square statistic, a goodness-of-fit measure, is 88.06% and the mean square root error is 0.52%. It is also possible to compute the higher-order model (e.g., cubic fitting) but the second-order one provides a sufficient level of insights of each parameter on the station performance. Some key quantities such as the Jacobian (Equation 7.7) and the Hessian matrix (Equation 7.8) are given to aid assessing the sensitivity of P_B with respect to C, B, ν, and λ. It can be seen that grid power C has the highest impact on decreasing the blocking probability, which is consistent with the simulation results.

$$
\begin{bmatrix} \frac{\partial P_B}{\partial C} \\ \frac{\partial P_B}{\partial B} \\ \frac{\partial P_B}{\partial \nu} \\ \frac{\partial P_B}{\partial \lambda} \end{bmatrix} = \begin{bmatrix} -0.035C - 0.0645B - 0.002\nu + 0.21\lambda - 2.66 \\ 0.0014C + 0.11B - 0.008\nu + 0.094\lambda - 1.62 \\ -0.002C - 0.078B + 0.178\nu + 0.025\lambda - 0.15 \\ 0.209C + 0.094B + 0.003\nu - 0.54\lambda + 3.84 \end{bmatrix} \qquad (7.7)
$$

7.7 Chapter Summary

This chapter proposed a fast-charging station architecture that is designed to protect the power grid by controlling the number of customers to be admitted for service at any given time. The critical component of the station design is the energy storage

$$H = \begin{bmatrix} -0.035 & -0.0645 & -0.002 & 0.21 \\ 0.0014 & 0.11 & -0.008 & 0.094 \\ -0.002 & -0.078 & 0.178 & 0.025 \\ 0.209 & 0.094 & 0.003 & -0.54 \end{bmatrix} \tag{7.8}$$

unit. We assessed the effect of including storage units into the analysis of system performance (i.e., service blocking probabilities). Moreover, we provided a simple financial model to develop guidelines to select ESS under different traffic regimes. It is noteworthy that the proposed system is designed to calculate the system resource for the peak time period. Therefore, the ESS can be used to participate in real-time energy markets to increase the financial profits. Furthermore, we showed that different energy storage technologies with faster charging and discharging rates could improve system performance. Therefore, we introduced another charging strategy, in which charging is first served by the ESS, and showed that if the local ESS has a high power rating, the station can charge more vehicles. The last part of the chapter employed response surface methodology, which is a statistical tool to show the relationship between the response parameter, the blocking probability, the system inputs grid power, ESS size, and the ESS charging rate in a polynomial form.

References

[1] Bayram, I.S., et al., "Electric Power Allocation in a Network of Fast Charging Stations," *IEEE Journal on Selected Areas in Communications*, Vol. 31, No. 7, 2013, pp.1235-1246.

[2] *fastned.nlen*

[3] Bai, S., Yu, D., and Lukic, S., "Optimum Design of an EV/PHEV Charging Station with DC Bus and Storage System," *Proc. IEEE Energy Conversion Congress and Exposition*, Atlanta, GA, Sep. 2010, pp. 1178-1184.

[4] Aggeler, D., et al., "Ultra-fast DC-charge Infrastructures for EV-mobility and Future Smart Grids," *Proc. IEEE Innovative Smart Grid Technologies Conference Europe*, Gothenburg, Sweden, Oct., 2010, pp. 1-8.

[5] Bayram, I.S., et al., "Local Energy Storage Sizing in Plug-in Hybrid Electric Vehicle Charging Stations under Blocking Probability Constraints," *Proc. IEEE International Conference on Smart Grid Communications*, Brussels, Belgium, Oct. 2011, pp. 78-83.

[6] Bayram, I.S., et al., "Strategies for Competing Energy Storage Technologies for in DC Fast Charging Stations," *Proc. IEEE International Conference on Smart Grid Communications*, Tainan City, Taiwan, Nov. 2012, pp. 1-6.

[7] De Freige, M., et al., "Power & Energy Ratings Optimization in a Fast-Charging Station for PHEV Batteries," *Proc. IEEE Electric Machines & Drives Conference*, Niagara Falls, ON, Canada, May 2011, pp. 486-489.

[8] Joos, G., et al., "Design and Simulation of a Fast Charging Station for PHEV/EV Batteries," *Proc. IEEE Electrical Power and Energy Conference*, Halifax, NS, Canada, Aug. 2010, pp.1-5.

[9] Bahrami, S. and Parniani, M., "Game Theoretic based Charging Strategy for Plug-in Hybrid Electric Vehicles," *IEEE Transactions on Smart Grid*, Vol. 5, No. 5, pp.2368-2375.

[10] Myers, R.H., Montgomery, D.C. and Anderson-Cook, C.M., *Response Surface Methodology: Process and Product Optimization Using Designed Experiments*, Fourth Edition, Hoboken, NJ: John Wiley & Sons.

Chapter 8

PEV Demand Control

8.1 Introduction

The last two chapters provided large- and small-scale charging station models, in which the primary goal was to solve the capacity planning and energy storage sizing problems for addressing a *given* traffic regime (e.g., peak hour of the day/month) and average customer demand. These problems have static forms in which the values of the decision variables (i.e., grid power and energy storage size) do not vary over time or in response to variations in the system state (e.g., the number of vehicles in the station). We refer to the problems given in Chapters 6 and 7 as static control or open-loop control problems. In fact, the planning decisions associated with such stations are generally carried out for longer time periods. Hence, upgrading the system through, for instance, incorporating new storage units or renewable sources (illustrated in Figure 8.1) are undesirable by the system operators as such upgrades tend to delay the return on investment.

While static control settings can address the current demands effectively, the customer demands tend to gradually increase due to growing PEV adoption and infrastructure-related issues, such as charging stations experiencing temporarily high demand due to road closures and detours. In such circumstances, a station operator needs to take dynamic actions to match the demand with the available resources. What distinguishes the dynamic design problems, which is the topic of this chapter and the following one, is that some of the station parameters are subject to dynamic decisions, where such decision actions are made with explicit attention to economic interpretations based on the guiding principles of the station operator

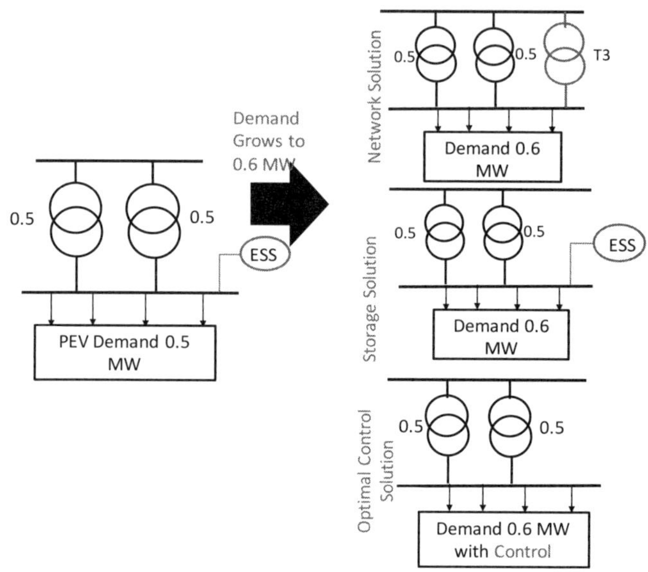

Figure 8.1 Solutions for managing growing demand.

[1]. The distinctive components of the design for dynamic settings including the following items:

- *Decision variables:* Customer arrival rates and vehicle charging power are included decision variables.
- *Costs and benefits:* Under dynamic settings the customers face costs associated with being declined service by a station and waiting to be served, while at the same time enjoy benefits associated with service completion and receiving incentives (e.g., reduced cost of service).
- *The objective:* Costs and benefits can be traded off to strike an optimal balance. This can be facilitated by adopting a proper *utility* function and optimizing it based on the network preferences and resources. For instance, each individual can seek to optimize its utility (*individually*), or in the settings in which there are multiple classes of customers (e.g., fast-charging and slow-charging), an agent can maximize the utility of a *class* as a whole. Additionally, a system operator might be interested in maximizing the aggregate

utility of all PEVs (*social-optimality*) or in maximizing stations' net profit (*station-optimality*).

In this chapter, we provide two pricing-based arrival control methods. In the first one, we consider a *single* fast charging station and aim to provide sufficient incentives in order to be able to shift the demand in time. This model seeks to maximize stations' profit. In the second model we consider coexistence of multiple PEV charging technologies and develop a pricing mechanism to control PEV demand in order to maximize social welfare.

8.2 Load Management for Fast-Charging Stations

The first control model considers a single fast-charging station. Efficient operation of such facilities in peak hours, during which customer demands rise, constitutes the bottleneck of the design. The load requested from the station is evidently shaped by customer preferences, and the fundamental operation of the charging station is primarily governed by service prices and incentives. Recall from the previous chapters that stations are often designed with fixed capacities. In such circumstances, economic incentives in the form of discounted service prices can be leveraged to align the demand with the station resources.

8.2.1 Load Management at a Fast-Charging Station

We consider a facility that provides service to *pass-by* customers who seek to receive the charging service with minimal delay. We consider a time-slotted service in which services are provided in periods $\mathcal{K} \triangleq \{1, 2, \ldots, K\}$, where the $k\text{th}$ slot has duration T_k. For customers, we use the same set of statistical assumptions that we used in the previous chapters. Specifically, the customers are assumed to arrive at the charging station following a Poisson process with the arrival rate λ_k during period k. Similarly, the charging service follows an exponential distribution with an average duration $1/\mu$ [6]. The charging service price is an exogenous variable, and hence, it is set in advance and is made available to all customers. Due to technological constraints (e.g., transformer and line ratings), the power grid can serve a maximum of C PEV users simultaneously at any period k. Hence, the corresponding system capacity is $50 \times P_{\text{power}}$ kW if the station is equipped with P_{power} kW chargers. To keep the focus squarely, we assume that vehicles are only charged by the grid power. As a result, the system can be modeled as an $M/M/C/C$ queue and the service blocking-probability at period k is given by

$$\beta_k (C, \rho_k) = \frac{\rho_k^C / C!}{\sum_{i=1}^{C} \rho_k^C / C!}, \tag{8.1}$$

where $\rho_k \lambda_k / \mu$ represents the traffic intensity during period k. Under heavy traffic regimes, the station operator may exert high service-blocking. From the station operator's point of view, it is desirable to regulate the arrival of PEV charging requests to guarantee a service blocking probability ε that is acceptable to the customers and also optimizes the distribution of the service over time. The optimal arrival rate, which can satisfy the target service- blocking probability, is given by [1]

$$\lambda^* \leq \arg\max_{\lambda} (\beta (C, \rho) \leq \varepsilon) . \tag{8.2}$$

In order to regulate customer demands so that they comply with λ^* during high demand periods, the charging station offers a discount to PEV users who postpone their charging request by one period.[1] An example of the intended traffic regulation is illustrated in Figure 8.2. In this example, during periods 2 and 3, the arrival rates of the charging requests exceed the optimal rate λ^*. Hence, PEV drivers are asked to delay their service for one period. Hence, the charging requests delayed from periods 2, 3, and 4 are served during periods 3, 4, and 5, respectively. It is assumed that the charging requests of PEV users will be postponed at most by one period. The PEV users who accept the discounted service rate and postpone their charging requests by one period will be admitted to the charging service in the next period. The customers who decline deferring their requests and demand immediate service take the risk of being subject to service blocking.

Each PEV user is willing to pay a units of service fee if it receives its service immediately and is willing to pay a_d if it defers its service and receives discount d_k. If a customer agrees to defer its service, therefore, it should satisfy $d_k > a - a_d$. For the aggregate behavior of PEV users, the values of a and a_d can vary randomly. Therefore, we define A and A_d as random variables to capture the random nature of a and a_d, respectively. Thus, for any incoming charging request, a customer is willing to pay a random amount $A \in [p, \infty]$ for immediate service and $A_d \in [0, a)$ for the deferred service. Hence, the probability that a discount offer made by the charging station is accepted by a PEV user with charging request is

$$F(d) = \mathbb{P}\{A - A_d \leq d\} . \tag{8.3}$$

1 Since we aim to shift the peak charging requests to an off-peak period, the period duration is set based on how long the peak duration lasts.

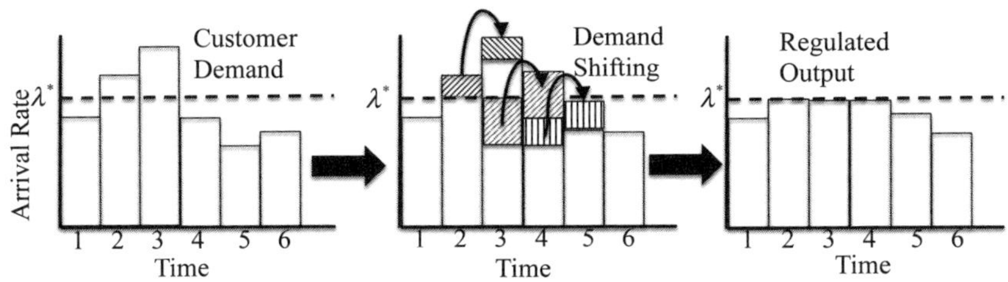

Figure 8.2 System model for PEV users.

The offered discount is denoted by d_k at period k and the facility revenue $R_k(d_k)$ can be shown as

$$R_k^U(d_k) = \lambda_k T_k(1 - \beta_k(C, \lambda_k))(p - d_k)F(d_k) + p(1 - F(d_k)) \, .$$

In the above equation, the first term corresponds to the users who accept the discount, while the second component corresponds to the users who request immediate service. Hence the combined services requested during the $k\text{th}$ period and those deffered from the $(k-1)\text{th}$ period should satisfy

$$\lambda_k(1 - F(d_k)) + \lambda_{k-1}F_{k-1}(d_{k-1}) \leq \lambda^* \, . \tag{8.4}$$

Based on this framework, the station operator seeks to maximize its own revenue by calculating the minimum amount of discount, which is given by

$$\begin{aligned} \max_{d_k > 0} \quad & R_k^U(d_k) \\ \text{s.t.} \quad & (8.4). \end{aligned} \tag{8.5}$$

It can bc readily verified that the optimal choice of the discount value, denoted by d_k^*, satisfies the optimal rate constraint with equality during the demand periods. Moreover, during low traffic regimes (i.e., $\lambda_k \leq \lambda^*$) the station operator does not offer any discounts, as it can accommodate all requested service. Therefore, the optimal discount rate is given by

$$d_k^* = \begin{cases} F_k^{-1}\left(\frac{\lambda_k + \lambda_{k-1} - \lambda^*}{\lambda_k}\right), & \lambda_k + \lambda_{k-1} > \lambda^* \\ 0, & \text{otherwise} \end{cases} \, . \tag{8.6}$$

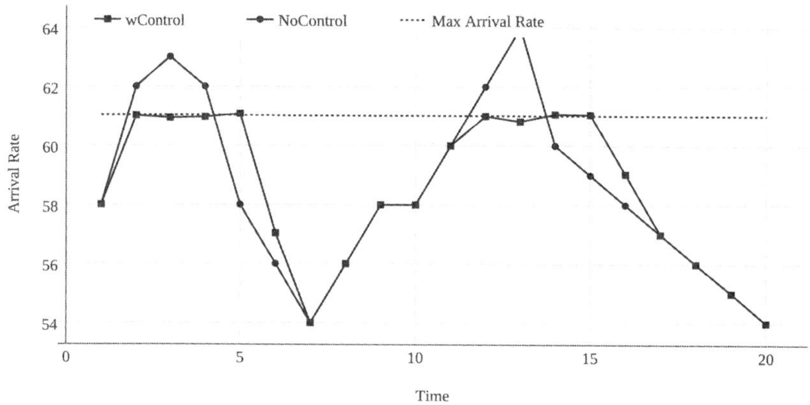

Figure 8.3 Customer arrival shaping.

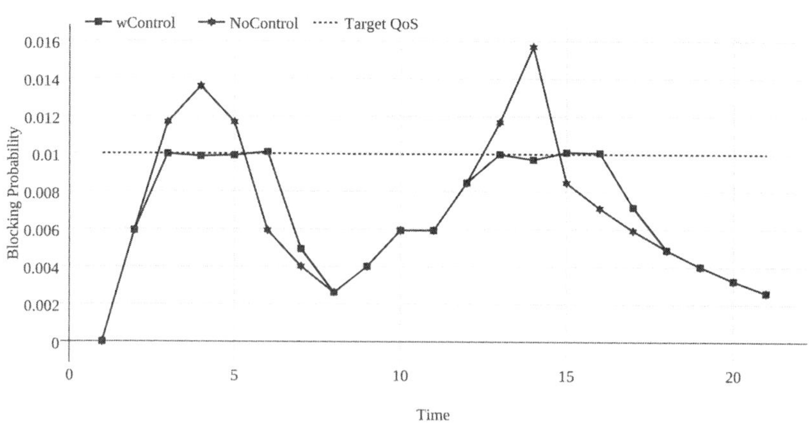

Figure 8.4 Peak load reduction.

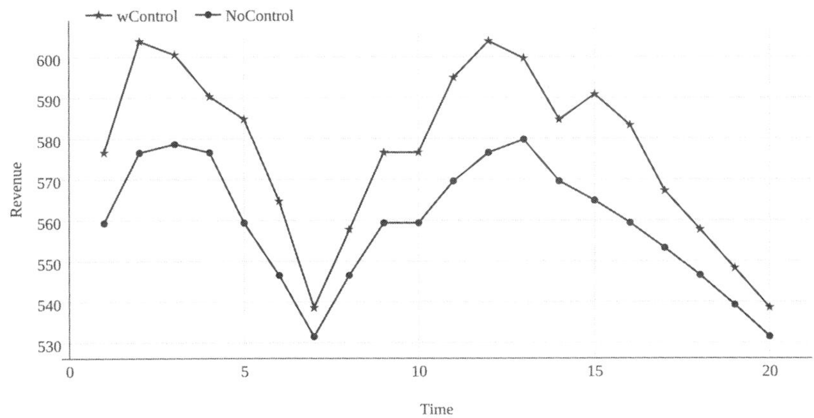

Figure 8.5 Revenue comparison.

8.2.2 Numerical Evaluation

We consider a fast-charging station with a total capacity of 1.5 MW and $C = 30$ physical chargers. The system parameters are set as follows. The duration of a time slot is set to one hour, and it is assumed that on average a fast charger can charge a PEV battery in 20 minutes ($\mu = 3$). The station operator sets the target blocking probability as $\epsilon = 0.01$, hence, the critical arrival rate becomes $\lambda^* = 61.02$. When the arrival rate is less than or equal to λ^*, the price for the charging service is 10 USD. On the other hand, if the customer arrival rate exceeds λ^*, the station operator offers discounts to *shape* the customer demand. Furthermore, the charging service fee that the customers are willing to pay follows a truncated Gaussian distribution with mean $m = 15$ and standard deviation $\sigma = 5$. Similarly, the mean and standard deviation of the service fee the customers are willing to pay for a deferred service are $m_d = 5$ and $\sigma_d = 3$.

The station performance is evaluated under varying arrival rates. To incorporate human activity, it is assumed that the station experiences morning and afternoon peak demands. Figure 8.3 compares the arrival rates of the settings with and without pricing-based control mechanism. Similarly, the blocking probability performance is shown in Figure 8.4. It is observed that customer demands can be shifted to meet

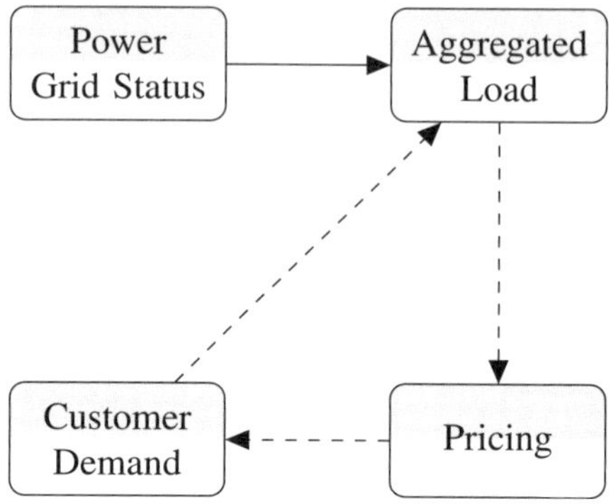

Figure 8.6 Incentive-based control mechanism for social welfare maximization.

the QoS targets, which not only boosts customer satisfaction but also increases the station's profit. Profit comparisons are shown in Figure 8.5.

8.3 Load Management at Multiple Charging Technology

In this section, we focus on a PEV charging network equipped with different charging technologies, such as level 1, level 2, and fast-charging. Similar to the previous case, the goal is to propose a control framework to regulate PEV demands and prevent system failures. Moreover, the framework leverages pricing dynamics in order to control customer demand and to provide a charging service with a good level of QoS. The overall framework is depicted in Figure 8.6.

The objective of the proposed framework is to incentivize and control the PEVs to submit their charging demands at certain arrival rates to maximize the social welfare. Such a framework assumes that the charging network is located in large metropolitan areas in which the PEV demand is highly stochastic and large. Next, we provide the details of the mathematical model.

8.3.1 System Model

We consider a facility that provides charging service to N customers and J distinct classes, labeled by $j \in \{1, .., J\}$. The customers are categorized based on their preferences; the size of battery packs, the amount of requested demand, and the charging rates denoted by $\{b_1, \ldots, b_J\}$. Since the station resources are limited by the distributed grid constraints, upon the arrival of a PEV type j, if the available station resources are less than b_j the customer will not be accepted to be served. Hence, the probability of being rejected (interchangeably used with loss-of-load-probability) arises as a natural performance metric. Due to the temporal fluctuations on the power grid, the amount of station resources varies over time. Thus, we consider a dynamic system represented by the time index $k \in \mathbb{N}$ in which C_k denotes the aggregate units of grid power available to the entire PEV fleet at time k. Note that C_k is an exogenous parameter given to the stations. Since the charging stations reside in a small well-confined region, the set of PEV chargers can be abstracted collectively as a superstation serving different customer classes. We further assume that system capacity is larger than the aggregate demand of each customer type.

An increase in the rate of the outage events penalizes the profit of the system operator and leads to service disruptions for the vehicles. Therefore, the optimal operation of the charging network is strongly linked to devising a set of decision rules for managing loss-of-load-probabilities (LoLPs), defined as the likelihoods that station resources fall below the aggregate customer demand [3]. The leverage that the system operators have for tuning the LoLP rates at the desired rates is the dynamically varying pricing scheme, by which they can influence the behavior patterns of the customers, which are abstracted by their arrival rates. To formalize this, we define $p_j(k)$ as the price of service of type j and define the price vector $\boldsymbol{p}(k) \triangleq [p_1(k), \ldots, p_J(k)]$, accordingly.

Based on the prices, which are known to all customers, each customer group decides whether to generate a service request of certain type. Let us denote the rate of type $j \in \{1, .., J\}$ requests generated by customer $n \in \{1, .., N\}$ during period k by $\lambda_j^n(k; \boldsymbol{p}(k))$. Hence, the aggregate arrival rate of service requests of type j is

$$\lambda_j(k; \boldsymbol{p}(k)) \triangleq \sum_{n=1}^{N} \lambda_j^n(k; \boldsymbol{p}(k)) \,. \tag{8.7}$$

From the network operator's point of view, these requests are being continuously generated across the networks, and as discussed in [6], a common assumption is that such aggregated requests (and not necessarily individual requests) arrive according to a Poisson process. Accordingly, we define the arrival rate vectors

$$\boldsymbol{\lambda}^n(k;\boldsymbol{p}(k)) \triangleq [\lambda_1^n(k;\boldsymbol{p}(k)),\ldots,\lambda_J^n(k;\boldsymbol{p}(k))] \,, \tag{8.8}$$

$$\text{and} \qquad \boldsymbol{\lambda}(k;\boldsymbol{p}(k)) \triangleq [\lambda_1(k;\boldsymbol{p}(k)),\ldots,\lambda_J(k;\boldsymbol{p}(k))] \,. \tag{8.9}$$

The set of feasible values for λ_j^n is defined in $[0,\infty)$. By noting that the LoLPs for each class are functions of the arrival rates, we define $\beta_j(k;\boldsymbol{\lambda}(k;\boldsymbol{p}(k))$ as the LoLP associated with class j, and we further define $\boldsymbol{\beta}(k;\boldsymbol{\lambda}(k;\boldsymbol{p}(k))$ as the corresponding LoLP vector. Finally, we follow the same assumption on the service rate statistics and define $1/\mu_j$ as the average charging duration for customer class j.

It is critical to note that since the charging stations reside in a small well-confined region such as a university campus, the entire group of chargers collectively act as one big station with multiple classes of customers. Therefore, power system losses are assumed to be negligible and all customers share the same resource pool. Moreover, for simplicity in the notation, in the rest of the discussions, the explicit dependence on k is omitted.

8.3.2 Problem Formulation

As stated earlier, the goal is to design a control framework that maximizes social welfare in the network of all customers. Hence, we define the utility function for PEV n as

$$U^n(\boldsymbol{\lambda}^n;\boldsymbol{\beta}(\boldsymbol{\lambda}))\,,$$

where the utility function depends on the arrival rate of customer n. Furthermore, due to having shared resources, the utility function also depends on the blocking probabilities associated with other classes. The aggregate utility in the system is

$$R = \sum_{n=1}^{N} U^n(\boldsymbol{\lambda}^n;\boldsymbol{\beta}(\boldsymbol{\lambda}))\,. \tag{8.10}$$

Hence, the goal of the system operator can be abstracted as maximizing (8.10) via optimally controlling the arrival rates $\{\boldsymbol{\lambda}^n\}_{n=1}^N$; that is,

$$\max_{\{\boldsymbol{\lambda}^1,\ldots,\boldsymbol{\lambda}^N\}} \sum_{n=1}^{N} U^n(\boldsymbol{\lambda}^n;\boldsymbol{\beta}(\boldsymbol{\lambda}))\,. \tag{8.11}$$

In the remainder of the analysis we assume that the utility function U^n is increasing in the arrival rates $\{\lambda_j\}_j$, and decreasing in the LoLPs $\{\beta_j\}_j$. We also assume that U^n is concave in λ_j for all $j \in \{1, \dots, J\}$, and continuously differentiable in all of its arguments.

8.3.3 Global Problem

The main goal of this section is to provide a methodology to solve the social welfare maximization problem given in (8.11). Since the global objective function depends on the utility functions of all customers, this problem is referred as the *global problem*. It is, however, noteworthy that PEVs are autonomous entities and operate based on common information known to all customers (e.g., charging prices) and their local perception about network dynamics. Such a distributed nature necessitates solving (8.10) in a distributed manner.[2] Therefore, the customers adjust their demands dynamically only based on the updated prices set by station operator. We show that there exist such distributed adjustments that lead to maximizing the social welfare given in (8.11).

Recall that individual utility functions U^n are increasing in the arrival rates $\{\lambda_j\}_j$, decreasing in the LoLP $\{\beta_j\}_j$, concave in $\{\lambda_j\}_j$, and continuously differentiable in all of its arguments. Then, the aggregate social welfare $R = \sum_{n=1}^{N} U^n$ is maximized when $\forall n \in \{1, \dots, N\}$ and $\forall j \in \{1, \dots, J\}$ we have

$$\frac{\partial R}{\partial \lambda_j^n} = \frac{\partial U^n}{\partial \lambda_j^n} + \sum_{l=1}^{N} \sum_{s=1}^{J} \frac{\partial U^l}{\partial \beta_s} \cdot \frac{\partial \beta_s}{\partial \lambda_j^n} = 0 \,. \tag{8.12}$$

By taking into account that the blocking probability of each customer class depends only on the sum of the arrival rates of that type; that is $\lambda_j = \sum_{n=1}^{N} \lambda_j^n$, from (8.12) we immediately have $\forall n \in \{1, \dots, N\}$ and $\forall j \in \{1, \dots, J\}$

$$\frac{\partial R}{\partial \lambda_j^n} = \frac{\partial U^n}{\partial \lambda_j^n} + \sum_{l=1}^{N} \sum_{s=1}^{J} \frac{\partial U^l}{\partial \beta_s} \cdot \frac{\partial \beta_s}{\partial \lambda_j} = 0 \,. \tag{8.13}$$

Therefore, solving (8.13) yields a *globally* optimal set of arrival rates for the entire customer population.

2 The problem could be solved in a centralized fashion if the vehicles were part of one coordinated fleet that is obligated to follow the instructions of a central authority.

8.3.4 Local Problems

The previous subsection shows that the global optimization solutions yielded by (8.13) can be calculated in a distributed manner, in which each charging station forms and solves a local problem. In order to solve this global problem and formalize such dynamic between posted prices and adjusted arrival rates, by recalling that p_j is the price charged to customers of type j, we first define the following local utility function for each customer $n \in \{1, \dots, N\}$ and service type $j \in \{1, \dots, J\}$.

$$\tilde{U}^n(\boldsymbol{\lambda}^n; \boldsymbol{\beta}(\boldsymbol{\lambda})) \triangleq \underbrace{U^n(\boldsymbol{\lambda}^n; \boldsymbol{\beta}(\boldsymbol{\lambda}))}_{\text{gain}} - \underbrace{\sum_{l=1}^{J} p_l \lambda_l^n (1 - \beta_l)}_{\text{cost}} . \tag{8.14}$$

By solving this local optimization problem over the relevant arrival rates, customer n computes its locally optimal and relevant arrival rates $\boldsymbol{\lambda}^n = [\lambda_1^n, \dots, \lambda_J^n]$. Specifically, the locally optimal arrival rates that maximize the local utility functions $\{\tilde{U}^n\}_n$ for all $\forall n \in \{1, \dots, N\}$ and $j \in \{1, \dots, J\}$ satisfy

$$\frac{\partial U^n}{\partial \lambda_j^n} - p_j(1 - \beta_j) = 0 . \tag{8.15}$$

By solving (8.15) each customer n can compute its own arrival rates based on the announced prices. The composition of the optimal local arrival rates depends on the choice of the utility function U^n.

8.3.5 Connection between Global and Local Problems

It is noteworthy that individual PEVs do not have any information about the derivatives of their arrival rates with respect to the set of LoLPs; that is, i.e., $\{\beta_j\}_j$ [4]. Also, the size of the system is large when compared to the demand of each individual customer. Hence, each customer does not have a significant impact on the LoLP. Therefore, by comparing equations (8.13) and (8.15) we can conclude that the optimal price for each $j \in \{1, \dots, J\}$ satisfies

$$p_j^* = -(1 - \beta^j)^{-1} \sum_{l=1}^{N} \sum_{s=1}^{J} \frac{\partial U^l}{\partial \beta^s} \cdot \frac{\partial \beta^s}{\partial \lambda_j} = 0 . \tag{8.16}$$

When a PEV of type j is presented with price p_j^*, the solution of its local optimization problem (8.14) satisfies the conditions given in (8.12), which in turn guarantee that the solution of the global optimization problem in (8.13) is equivalent to the local ones given in (8.14) as established in [4]. Hence, the local optimal solutions become the equilibrium solutions since no single customer can obtain a higher utility by deviating from its locally optimal solution.

8.3.6 Computing LoLPs

The station model described in the previous section falls into multidimensional loss systems (or multirate Erlang loss systems), which are employed to assess the guaranteed performance of networks with finite resources. In such systems, an arriving customer requesting a certain amount of resources is either accepted for service or is rejected. Moreover, the utility functions depend on the system-blocking probabilities and their derivatives. Therefore, we are interested in computing the LoLP as a function of system parameters.

We first describe a toy example to illustrate the intuition behind the multidimensional system model. Suppose that there are three customer classes requesting $j = \{2, 3, 4\}$ amount of resources, and the station capacity is $C = 8$. We are interested in computing blocking probabilities for each class, which is the probability that an arriving PEV will not find enough resources. In this case, the system state is three-dimensional and represents the active number of customers of each type. In this example, there can be at most four customers of type 1, two customers of type 2, and one customer of type 3. Computing blocking probabilities are complicated as the station could serve a mix of different customer types. In this case, the LoLP-related computations need to include all of the *blocking states*. In this toy example, for customer type 3 the blocking occurs when the amount of available resources is less than three units. Hence, the blocking probability pertinent to type 3 is the sum of the steady-state probabilities $\pi(4,0,0) + \pi(3,0,0) + \pi(2,1,0) + \pi(2,1,0) + \pi(2,0,1) + \pi(1,0,1) + \pi(1,1,0) + \pi(1,2,0) + \pi(0,2,0) + \pi(0,1,1) + \pi(0,0,2)$.

This toy example shows that the system size grows exponentially as the number of classes and station capacity increases. Therefore, efficient computation of the LoLP requires analyzing J independent time-reversible Markov chains. As mentioned above, the state of the system is defined as the number of customers of each type, that is $\boldsymbol{Q} \triangleq [Q_C^1, \ldots, Q_C^J]$ and the state space is denoted by $\Omega \triangleq \{\boldsymbol{Q} : \sum_{j=1}^{J} b_j Q_\infty^j \leq C\}$. Let $\tilde{Q}_C^j$ denote the maximum number of customers of type j that can be served simultaneously. Assuming the order $b_1 \geq \ldots \geq b_J \geq 0$, without loss of generality, provides that $0 \leq \tilde{Q}_C^1 \leq \ldots \leq \tilde{Q}_C^J$. Then the probability of being at state $\boldsymbol{Q}$ is [5]

$$\overline{\pi}(\mathbf{Q}) = \prod_{j=1}^{J} \frac{q_j^{Q_\infty^j}}{Q_\infty^j!} e^{-q_j} \ . \tag{8.17}$$

For a given set of charging rates $\boldsymbol{b} \triangleq [b_1, \ldots, b_J]$ and average numbers of arrival rates $\boldsymbol{q} \triangleq [q_1, \ldots, q_J]$, we denote the LoLP by of customer type j by

$$\beta_j : \mathbb{R}^J \to [0,1] \ ,$$

which captures the connection between the block probability on one hand, and the customer statistics $\boldsymbol{b}$ and $\boldsymbol{q}$, on the other hand. By noting that the numbers of PEVs for different classes Q_∞^j are mutually independent Poisson random variables, we have

$$\beta_j(\boldsymbol{q}, \boldsymbol{b}) = \mathbb{P}\left\{ C - b_j < \sum_{j=1}^{J} b_j Q_C^j \right\} \tag{8.18}$$

$$= \mathbb{P}\left\{ C - b_j < \sum_{j=1}^{J} b_j Q_\infty^j \leq C \mid \sum_{j=1}^{J} b_j Q_\infty^j \leq C \right\} \tag{8.19}$$

$$= \frac{\mathbb{P}\left\{ C - b_j < \sum_{j=1}^{J} b_j Q_\infty^j \leq C \right\}}{\mathbb{P}\left\{ \sum_{j=1}^{J} b_j Q_\infty^j \leq C \right\}} \ . \tag{8.20}$$

Similar to (8.19), we condition on the finite capacity and compute a generic state $\boldsymbol{Q}$ probability distribution as

$$\pi(\boldsymbol{Q}) = \frac{\overline{\pi}(\mathbf{Q})}{\sum_{\tilde{Q} \in \Omega} \overline{\pi}(\tilde{\mathbf{Q}})} \cdot \tag{8.21}$$

By recalling the discussion given in the toy example, we define the *blocking states* for customer type j as

$$\Psi_j \triangleq \{ \boldsymbol{Q} : C - b_j \ < \ \sum_{k=1}^{J} b_k Q_C^k \ \leq \ C \} \ .$$

Hence, (8.20) can be rewritten as:

$$\beta^j(\boldsymbol{q}, \boldsymbol{b}) = \sum_{s \in \Psi_j} \pi(s) = 1 - \sum_{s \notin \Psi_j} \pi(s) \,, \tag{8.22}$$

where the second term denotes the probability that the charging station has a total capacity of $(C - b_j)$ instead of C, and $\pi(s)$ is the steady state probability mass function. Moreover, we define function $H(C, J)$ as

$$H(C, J) \triangleq \sum_{\{\boldsymbol{Q}\colon \boldsymbol{bQ} \leq C\}} \prod\nolimits_{j=1}^{J} \frac{q_j^{Q^j}}{Q^j!} \,, \tag{8.23}$$

based on which the individual LoLPs can be computed according to

$$\beta_j(\boldsymbol{q}, \boldsymbol{b}) = 1 - \frac{H(C - b_j, J)}{H(C, J)} \,. \tag{8.24}$$

The set $\{\boldsymbol{Q} : \boldsymbol{bQ} \leq C\}$ in (8.23) contains all the system states corresponding to a situation in which no blocking events occur. While (8.24) provides an explicit representation for β_j, the associated computations can be costly when the system capacity C is large. Considering a real-world situation in which the number of classes J typically varies between three and six and the station capacity is in the order of Mega-watts, calculating the LoLPs can be carried out via the Kaufman-Roberts algorithm (Algorithm 1), which involves a simple recursion for computing the occupied resources [6]. Let c denote the amount of resources being in use and define

$$\alpha(c) \triangleq \mathbb{P}\{\text{c units of power in use}\} \,.$$

Hence, we have

$$\alpha(c) = \sum_{\{\boldsymbol{Q}\colon \boldsymbol{bQ} \leq c\}} \frac{q_j^{Q^j}}{Q^j} \cdot \frac{1}{H(C, J)} \,, \tag{8.25}$$

and the LoLP corresponding to customer type j is given by

$$\beta_j(\boldsymbol{q}, \boldsymbol{b}) = \sum_{i=0}^{b_j - 1} \alpha(C - i) \,. \tag{8.26}$$

Notice that the above derivations are based on the premise that resources supplied from the grid are discretized (e.g., 1 kW is considered as 1000 discrete serving units).

Recall that the global and the local optimization problems contain the set of derivatives of LoLP with respect to offered load for each PEV class. In the proposed multi-class customer model, there is no explicit formula for the performance measures (LoLP) in terms of input parameters. In order to efficiently find the derivatives, we follow the methods based on convolution algorithms [7] . To this end, the derivative of the LoLP associated with customer type j with respect to the traffic intensity of another class $j_1 \neq j_2$ can be computed by using the function $\alpha(\cdot)$ as follows.

$$\begin{aligned}\frac{\partial \beta_{j_1}}{\partial q_{j_2}} &= \alpha(C - b_{j_1}) + \alpha(C - b_{j_2} - 1) + \cdots \\ &+\alpha(C - b_{j_2} - b_{j_1} - 1) - (1 - \beta^{j_2})\beta^{j_1} .\end{aligned} \tag{8.27}$$

An important property of LoLP calculations is the elasticity property, which shows that the sensitivity of customer j's LoLP to the traffic density of customer class k is the same as the sensitivity of customer k's LoLP to the traffic density of customer j [10]; that is,

$$\frac{\partial \beta_j}{\partial q_k} = \frac{\partial \beta_k}{\partial q_j} . \tag{8.28}$$

This property reduces the computational complexity, as the optimization problem contains such derivatives. Next, we provide an example with the following parameters. Assume that the charging network draws $C = 1000$ units of power from the grid and serves three types of customers (i.e., $j \in \{1, 2, 3\}$). Customer classes are differentiated by the charger technology they use, and by mimicking the current charging standards (e.g., fast-charging and level 2 charging) the charging statistics are set as $\boldsymbol{b} = \{50, 7, 5\}$, $\mu_1 = 3$, $\mu_2 = 0.42$, and $\mu_3 = 0.2$.

Algorithm 1 Kaufman-Roberts Algorithm [6]

Set $\kappa(0) = 0$ and $\kappa(i) = 0$ for $i \in \mathbb{R}^-$
for i=1 to C **do**
 $\kappa(i) = \frac{1}{i} \sum_{j=1}^{J} b_j q_j (j - b_j)$
end for
Compute $H = \sum_{i=1}^{C} \kappa(i)$
for i=0 to J **do**
 $\alpha(i) = \frac{\kappa(i)}{H}$
end for
for j=1 to J **do**
 $\beta_j(\boldsymbol{q}, \boldsymbol{b}) = \sum_{i=C-b_j+1}^{C} \alpha(i)$
end for

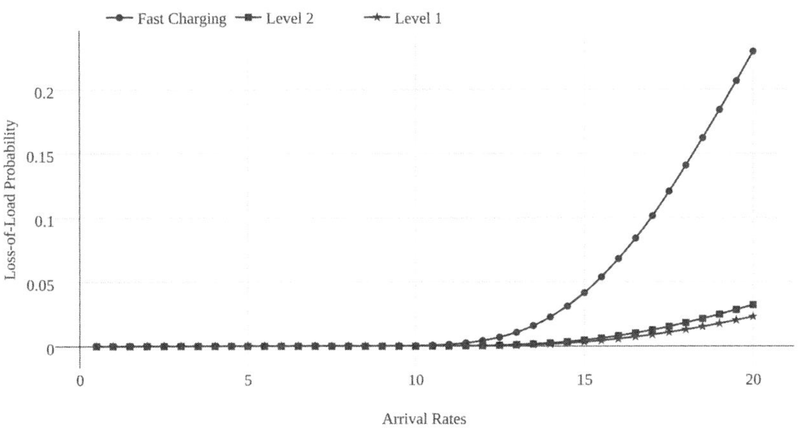

Figure 8.7 Loss-of-load probability calculation using Kaufman-Roberts algorithm for a 1MW charging station serving fast, level 2 and level 1 customers.

The results depicted in Figure 8.7 show that the blocking performance of the DC fast-charging customers is worse than those of the other classes. This is due to the fact that fast charging requires 50 kW of charging power, while the requirements for the rest of the classes are considerably lower.

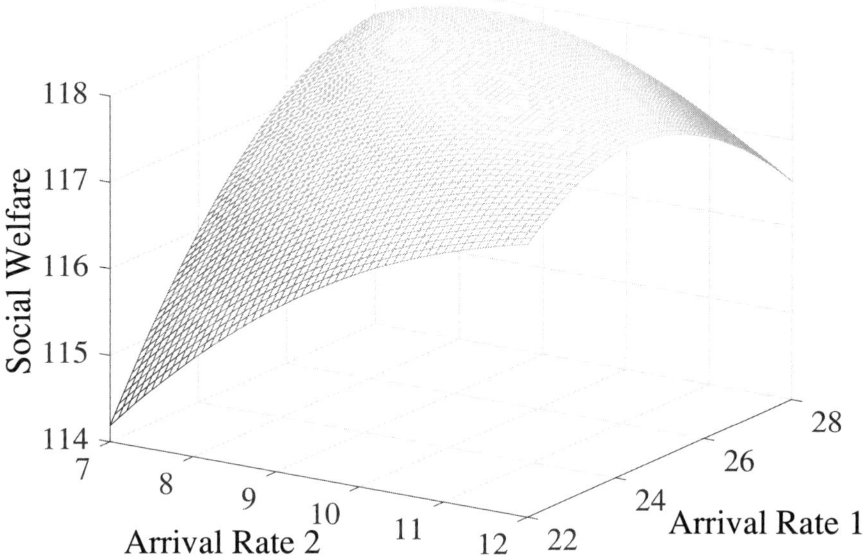

Figure 8.8 Social welfare optimization for 1 MW capacity.

8.3.7 Numerical Evaluation

In this section, we provide a case study to show how the pricing mechanism controls the customer demands and maximizes social welfare. We assume that there are two types of customers equipped with fast-charging and slow-charging technologies. Following the discussions in the previous section, the parameters are set as $b_1 = 50$, $b_2 = 7$, $\mu_1 = 3$, and $\mu_2 = 0.42$. For the utility function of a single user, we adopt the widely used logarithmic function in which the utility of a single customer increases with the arrival rate and decreases with the customer blocking rates. The utility function U, which is defined over $\lambda_j \geq 0$, is given by

$$U = \sum_{j=1}^{J} \omega_j \log(1 + \lambda_j) - \theta_j \log(1 + \beta_C^j) \, , \tag{8.29}$$

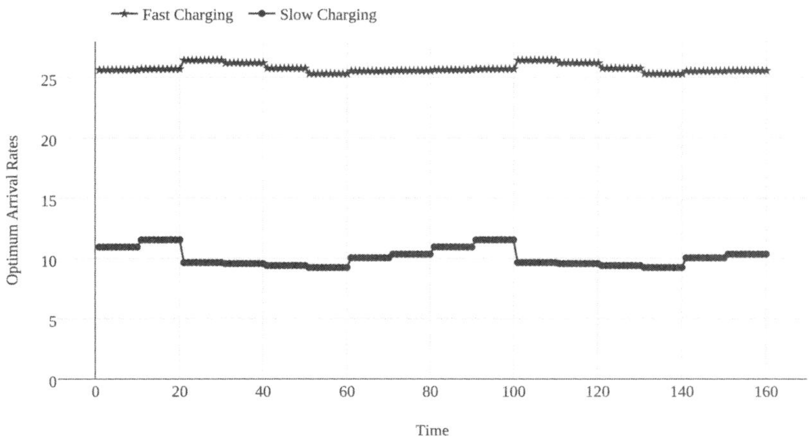

Figure 8.9 Optimal arrival rates.

where ω_j and θ_j are the weights of PEV class j. Note that the weights of the DC fast-charging customers are higher than those of the slow-charging ones because they enjoy short charging durations and gain more utility. To evaluate the effectiveness of the model, the station capacity is set to 1 MW, and other parameters are set as $\omega_1 = 30$, $\omega_2 = 10$. and $\theta_1 = 60$, and $\theta_2 = 20$. The optimal arrival rates can be obtained as $\lambda_1^* = 25.7$ and $\lambda_2^* = 9.4$, with blocking likelihoods of $\beta_1 = 0.006$ and $\beta_1 = 0.0007$. Based on these figures, the optimal prices become $p_1^* = 0.122$ and $p_2^* = 0.097$, with the aggregate utility being 117.55. To illustrate the variations of social welfare, we evaluate the system for varying arrival rates. The results in Figure 8.8 show that maximum welfare is attained at $(\lambda_1^*, \lambda_2^*) = (25.4, 9.4)$.

Next, we perform simulations for a wide range of system parameters. We assume that the station capacity varies over time and follows $C(k) = 1000 + 20\sin(2\pi k/80)$, and it is assumed to be constant for every $T = 10$ duration. Results depicted in Figure 8.9 show that it is more beneficial to accept fast-charging customers as they improve the social welfare function more than slow-charging ones. For the given set of arrival rates, the corresponding LoLPs are shown in Figure 8.10.

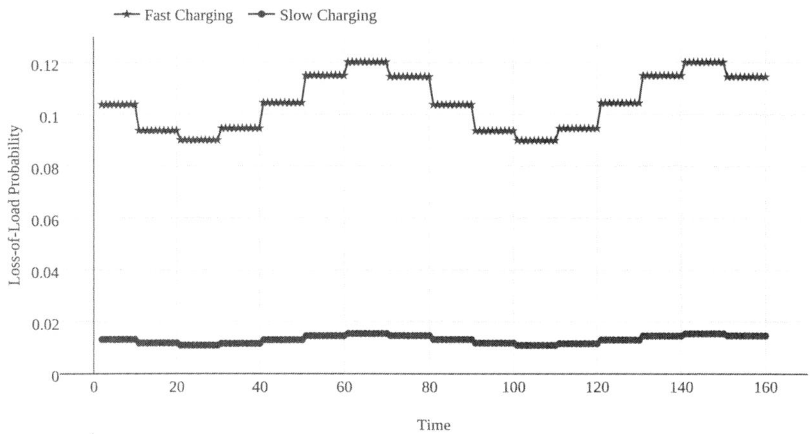

Figure 8.10 Load-of-loss probability.

8.4 Chapter Summary

This chapter provided pricing-based control frameworks for PEV charging infrastructures that are operating under heavy traffic regimes. The first framework is apt for small-scale DC fast charging stations (similar to the one discussed in Chapter 7), in which the aim of the station operator is to offer incentives to customers so that (1) the aggregate demand is kept below the station serving capacity, and (2) the station profit is maximized. The second pricing framework considered a more complex problem consisting of multiple classes of customers that share the grid resources, and the goal was to compute optimal arrival rates to maximize the social welfare of the system. The main distinctive component of the proposed models is the online interaction of station operator and PEV drivers. The simulations have shown that pricing could be employed as a tool to keep aggregate customer demand below station serving capacity.

References

[1] Stidham Jr, S., *Optimal Design of Queueing Systems*, CRC Press, 2009.

[2] Bayram, I.S., et al., "Electric Power Allocation in a Network of Fast Charging Stations," *IEEE Journal on Selected Areas in Communications*, Vol. 31, No. 7, 2013, pp.1235-1246.

[3] Von Meier, A., *Electric Power Systems: A Conceptual Introduction*, New York: John Wiley & Sons, 2006.

[4] Courcoubetis, C. A., & Reiman, M. I., "Pricing in a Large Single Link Loss System," *Proc. Proceedings of 16th International Teletraffic Congress*, Edinburgh, Jun. 1999.

[5] Kleinrock, L. (1975). *Queueing Systems, Volume 1: Theory*, New York: Wiley-Interscience, 1975.

[6] Kaufman, J. S., "Blocking in a Shared Resource Environment," *IEEE Transactions on Communications*, Vol 29, No. 10, 1981, pp. 1474-1481.

[7] Iversen, V. B. and Stepanov, S. N.,"Derivatives of Blocking Probabilities for Multi-service Loss Systems and Their Applications," In *Next Generation Teletraffic and Wired/Wireless Advanced Networking*, Koucheryavy, Y., Harju, J., and Sayenko, A. (eds.), Berlin. Springer, 2007, pp. 260-268.

[8] Bayram, I. S. et. al., "A Pricing-based Load Shifting Framework for EV Fast Charging Stations," In *IEEE International Conference on Smart Grid Communications*, Nov., 2014, pp. 680-685.

[9] Wang, G. et. al., "Dynamic Competitive Equilibria in Electricity Markets," in *Control and Optimization Methods for Electric Smart Grids*, New York: Springer, 2012.

[10] Mazumdar, R. R., "Performance Modeling, Loss Networks, and Statistical Multiplexing", *Synthesis Lectures on Communication Networks*, Vol. 2, No. 1, 2009, pp 1-151.

Chapter 9

Optimal Service Rates

9.1 Introduction

In the last three chapters, we provided a detailed analysis of charging station designs. More specifically, in Chapters 6 and 7, we provided mathematical frameworks for optimally provisioning the station resources such as power drawn from the grid and energy storage size. Furthermore, in Chapter 8 we considered the case in which customer population grows and station resources become inadequate to meet the demand. In this case, the station operator offered incentives to customers to shift their consumption to off-peak hours, so that the station, overall, can provide charging service with a desirable level of quality of service.

This chapter complements our work on single charging station design and takes the charging rate as the primary decision variable. The chapter is divided into two sections. The first section provides an analysis on optimally charging a single Li-ion battery. The results lead to the well-known constant current constant voltage (CCCV) charging regime. The second section, on the other hand, considers a charging station with arriving customers having different SoC levels. The primary motivation is that the charging power approaches zero when the SoC gets close to one hundred percent. Hence, by limiting the maximum SoC level, the station can serve more customers and maximizes its operating revenue, as the charging duration reduces significantly.

The charging rate-based control approaches consider a variety of competing performance measures. In the next section, we provide an overview of the measures that have gained popularity in the literature.

9.1.1 Competing Performance Measures for PEV Charging

Depending on the application and the scenario, PEV battery charging problems include a variety of different objectives such as electricity cost, power system losses, battery degradation, target load profile, and customer comfort. Next, we present an overview of each component.

- *Electricity cost* or the service fee is the first major objective of an optimization framework. For home or garage charging applications the electricity price is usually an exogenous variable given by the utility company, and the charging goal is to optimally adjust the charging rate to minimize the service cost [1, 2]. Recall that Section 4.2 presented an overview of pricing regimes, and a sample optimization framework was presented in Section 4.3 where the utility employs the time of use prices and the user adjusts the charging power. Pricing can be used as a control tool to manage PEV load, and two frameworks were introduced in Chapter 8.

- *Power system losses and voltage profiles* have been the focus of many studies, as PEVs represent significant loads that can impact power system performance, as discussed in Chapter 2. The charging and discharging of PEVs may also be done by aiming to minimize network losses [3–5] and the PEV battery can be used in optimal power flow problems. It is noteworthy that the studies in this direction usually consider home charging applications in which the duration for service completion is fairly long (e.g., 7-8 hours). Therefore, adjusting the charging rates does not affect customers preferences.

- *Battery degradation* is another important component affecting the charging rate. Chapter 4 provided a detailed analysis of battery degradation and provides several examples. Also, the charging station model takes battery degradation into account when determining service rates [1].

- *Customer satisfaction* is measured with QoS metrics such as the average waiting time for service, outage probability, and target SoC level. Stations with limited charging space (e.g., small-scale fast-charging stations) employ outage probability, while larger-scale stations with waiting space use both outage and waiting times as the performance metric. Moreover, customer satisfaction may include delaying the charging service or meeting a certain deadline (e.g., service completion by 7 am).

Note that the performance measures mentioned above, if considered concurrently in one framework, are likely to be competing. For instance, in order to reduce

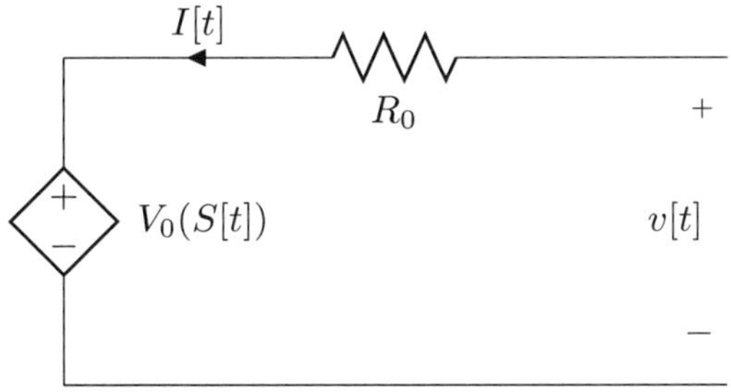

Figure 9.1 Circuit model of a PEV battery.

charging duration and maximize customer satisfaction, charging power should increase. However, this may degrade battery life, while the cost of service depends on time-of-use prices. Similarly, network conditions may limit the charging rates due to congestions and voltage sags. Therefore, the roles of the decision parameters are application-dependent. On the other hand, the model presented in this section considers electricity cost, energy losses, battery degradation, and customer satisfaction.

9.2 Optimal Individual Battery Charging

In this section, we analyze the optimal battery charging problem of a single PEV equipped with a Li-ion battery and provide the derivation of the widely accepted CCCV charging regime. The objective is to minimize the charging duration and energy losses related to energy transfer. To initiate our analysis, we present the simplified equivalent circuit model of a PEV battery in Figure 9.1. It is noteworthy that there are complex circuit models to capture the PEV battery operations [6]. Nevertheless, the equivalent circuit model presented in this chapter conveys enough details and allows us to derive the closed-form solution.

The elements of the circuit include a voltage-controlled voltage source and a resistor representing the internal resistance of the battery. For the voltage source, the corresponding open-circuit voltage (OCV) is denoted by V_0, which is a function of the battery state of charge (SoC), which is denoted by $S[t]$. Therefore, the OCV can be represented by $V_0(S[t])$. Clearly, the SoC is 1 when the battery is fully charged and it is 0 when the battery is entirely discharged. Moreover, the SoC is measured at discrete times, and where the samples are taken every δ seconds. Let us use t_0 and t_f to represent the initial and final times of charging, and similarly, $S[t_0]$ and $S[t_f]$

denote the initial and final SoC, which are assumed to be known. Furthermore, by setting $t_0 = 0$, the charging time of a PEV becomes $t_f\delta$[1]. Finally, it is assumed that the maximum allowed terminal voltage charging is v_m, and therefore, $v[t] \leq v_m$, $\forall t$.

At any intermediate charging time step t, the battery SoC follows

$$S[t+1] = S[t] + C_c i[t] \ , \tag{9.1}$$

where C_c is the coulomb counting parameter (in 1/Amperes). Since one coulomb per second is one ampere, C_c is represented by $C_c = \frac{\delta}{3600 C_{\text{bat}}}$, where C_{bat} is the battery capacity. As stated earlier, the goal of this section is to find the optimal charging pattern to minimize the charging duration (CD) and the energy losses (EL). The first part of the objective function, CD, can be represented by $F_{\text{CD}} = t_f\delta$. The second part accounts for the energy losses, which depend on the current flowing through the resistor. Therefore, the losses are proportional to the square of the current and the magnitude of the resistor, and for each time interval, it can be written in the following form[2]

$$F_{\text{EL}} = \sum_{t=0}^{t_f-1} R_0 (i[t])^2 \delta \ . \tag{9.2}$$

Then, the objective function becomes

$$F = \omega_{\text{CD}} F_{\text{CD}} + \omega_{\text{EL}} F_{\text{EL}} \tag{9.3}$$

$$= \omega_{\text{CD}} t_f \delta + \omega_{\text{EL}} \sum_{t=0}^{t_f-1} R_0 (i[t])^2 \delta \ , \tag{9.4}$$

where ω_{CD} and ω_{EL} are the weights of each cost component. Also, from the battery specifications R_0 is assumed to be known. It is noteworthy that only the ratio of the weights $\theta = \frac{\omega_1}{\omega_2}$ affects optimal current profile $i[t]$. Hence, the objective function in (9.4) can be written as

$$F = \theta t_f \delta + \sum_{t=0}^{t_f-1} R_0 (i[t])^2 \delta \ . \tag{9.5}$$

The optimal Li-ion battery charging behavior can be split into two parts. When the PEV battery is plugged in, the open-circuit voltage V_0 rises and pushes the

1 For the rest of the analysis, $t_0 = 0$ unless otherwise stated.

2 Notice that at $(t - t_f)$ the charging current becomes zero, hence the summation ends at $(t_f - 1)$.

terminal voltage to rise up to v_m, which is the voltage level when the battery is fully charged (i.e., $v_m = V_0$ and $S[t_f] = 1$). Typically, the terminal voltage reaches to v_m before the battery is fully charged. Let us assume that the terminal voltage reaches v_m at time t_k and it is kept constant until t_f. Then, the charging current during $t = t_k, t_{k+1}, \cdots t_{f-1}$ can be written as $i[t] = \frac{1}{R_0}(v_m - V_0(S[t]))$ and the SoC levels are $S[t_1] = S_1, \ldots, S[t_f] = S_{t_f}$. Therefore, the voltage and the current profiles are already found for $t \geq t_k$. Hence, the problem reduces to finding the charging profile until the terminal voltage reaches v_m during $t = 0, \cdots, t_k$. To that end, the optimal charging problem can be simplified to finding the profile until the voltage reaches v_m and can be written as

$$\begin{aligned} \text{minimize} \quad & \theta t_f \delta + \textstyle\sum_{t=0}^{t_f-1} R_0 (i[t])^2 \delta \\ \text{s.t.} \quad & S[t+1] = S[t] + C_c i[t] \\ & V_0(S[t]) + i[t] R_0 \leq v_m \\ & S[0] = S_0, \; S[t_k] = S_{t_k} \\ & 0 \leq S[t] \leq 1 \end{aligned} \quad . \tag{9.6}$$

The quadratic problem given above can be solved in three steps as described below [7–9]:

1. For a given t_k, calculate the optimal current profile, $i[t]$ for $0 \leq t \leq t_k$, so that it minimizes the energy losses and calculate F_{EL} in terms of t_k.

2. Construct a new objective function with the t_k-dependent minimum EL calculated from the first step, plus weight-charging duration cost. Then, calculate the optimal t_k based on the devised objective function.

3. Evaluate the optimal current profile obtained in step 1 with the optimal t_k attained in step 2.

The first step is evaluated as follows. For a given t_k, the optimal current profile can be obtained from minimizing the aggregate energy loss subject to the same constraints given in (9.6). The optimization problem can be solved by using the Hamiltonian function [10], which is given by

$$H[t] = R_0 i^2[t] \delta + \Upsilon[t+1](S[t] + C_c i[t]) \; , \tag{9.7}$$

Hence, for the optimal solution the following equalities must hold.

$$\Upsilon[t] = \frac{\partial H[t]}{\partial S[t]}, \tag{9.8}$$

$$0 = \frac{\partial H[t]}{\partial i[t]}, \tag{9.9}$$

$$\text{and} \qquad S[t+1] = \frac{\partial H[t]}{\partial \Upsilon[t+1]}. \tag{9.10}$$

Equation (9.8) leads to $\Upsilon[t] = \Upsilon[t+1]$ for $t \in \{0, 1, \dots, t_k - 1\}$ and $\Upsilon[t_k] = \nu$, where ν is the Lagrange multiplier of the constraint $S[t_k] = S_1$. We can further write that $\Upsilon[t] = \nu$ for $t \in \{0, 1, \dots, t_k - 1\}$. Moreover, from (9.9), the optimal current profile for $t \in \{0, 1, \dots, t_k - 1\}$ can be obtained as

$$i^*[t] = -\frac{C_c \Upsilon[t+1]}{2R_0\delta}, \quad \text{for} \quad t \in \{0, 1, \dots, t_k - 1\}. \tag{9.11}$$

By replacing $\Upsilon[t+1]$ with the Lagrange multiplier, the optimal current profile can be simplified to

$$i^*[t] = -\frac{C_c \nu}{2R_0\delta}, \quad \text{for} \quad t \in \{0, 1, \dots, t_k - 1\}. \tag{9.12}$$

Recall that the battery SoC follows $S[t+1] = S[t] + C_c i[t]$. If the initial battery SoC is known, the SoC at time t can be rewritten by inserting the optimal current given in (9.12). Hence,

$$S[t] = S[0] - \frac{t_k C_c^2 \nu}{2R_0\delta}. \tag{9.13}$$

Since the battery charging dynamics follow $S[t_k] = S_1$ at $t = t_k$, we can write $S_1 = S[0] - \frac{t_k C_c^2 \nu}{2R_0\delta}$ and calculate the Lagrange multiplier ν as

$$\nu = -\frac{2R_0\delta(S_1 - S[0])}{t_k C_c^2}, \tag{9.14}$$

Finally, plugging (9.14) into (9.12) yields

$$i^*[t] = \frac{S_1 - S[0]}{t_k C_c}, \quad \text{for} \quad t \in \{0, 1, \dots, t_k - 1\}. \tag{9.15}$$

Table 9.1
Summary of Optimal Battery Charging

Time	Voltage	Current
$0-(t_k-1)$	$V_0(S[t])$	$\sqrt{\frac{\theta}{R_0}}$
$t_k - t_f$	v_m	$\frac{1}{R_0}(v_m - V_0(S[t]))$

Step 1 of the solution process is finalized by writing the optimal energy loss cost function as

$$F^*_{\mathrm{EL}}(t_k) = \frac{R_0\delta(S_1 - S[0])^2}{t_k C_c^2} \,. \tag{9.16}$$

Having defined the energy loss function as a function of t_k – the time when the voltage reaches its maximum value v_m – the second step seeks to find the optimum t_k for the updated objective function

$$F = \theta t_k \delta + \frac{R_0\delta(S_1 - S[0])^2}{t_k C_c^2} \,. \tag{9.17}$$

The optimal t_k can be obtained by taking the derivative of (9.17) with respect to t_k and setting it equal to zero. Then, the optimal t_k^* becomes

$$t_k^* = \frac{S_1 - S[0]}{C_c}\sqrt{\frac{R_0}{\theta}} \,, \tag{9.18}$$

which finalizes step 2. In the final step, we simply plug (9.18) into (9.15), based on which the optimal current profile can be written as

$$i^*[t] = \frac{S_1 - S[0]}{t_k^* C_c} = \sqrt{\frac{\theta}{R_0}} \,, \quad \text{for} \quad t \in \{0, 1, \ldots, t_k - 1\} \,. \tag{9.19}$$

Note that the optimal current profile is constant and depends only on the weight of each cost component and the internal resistance of the battery. Therefore, (9.19) represents the constant current part of the CCCV charging regime. As mentioned before, during the constant-voltage part, the battery is charged at maximum allowable voltage; that is $V[t] = v_m$ for $t \in \{t_k, t_{k+1}, \cdots, t_f\}$. The summary of optimal charging profiles are shown in Table 9.1.

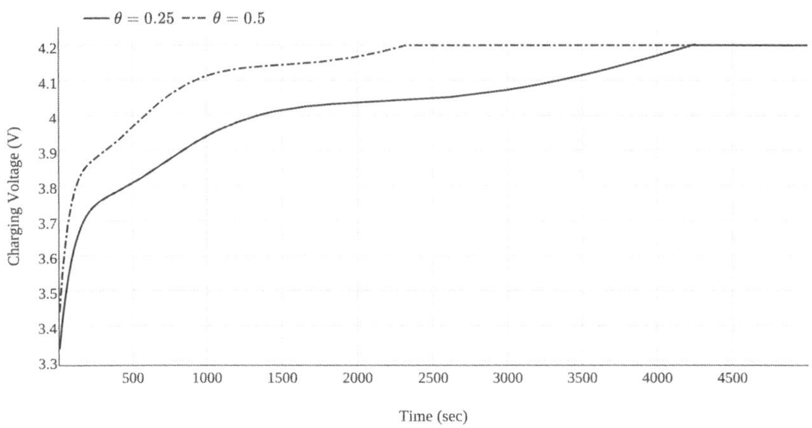

Figure 9.2 Optimal voltage profile.

9.2.1 Numerical Evaluation

Before proceeding to next section in which CCCV charging is applied in a charging station, we first provide a sample evaluation for a single Li-ion battery cell. The commercial battery model is BP-4L and the simulation parameters are $\delta = 1$ second, $S_0 = 0$, $S[t_f] = 1$, and $R_0 = 0.258$ ohm, the cell capacity is $C_{\text{bat}} = 1.5612$ Ah, and the θ is set to 0.25 and 0.50 for comparison of two cases. The details of the open-circuit voltage are represented by the following polynomial expression and the details can be found in [11] and [7],

$$V(k) = \sum_{i=0}^{4} A_i k^{-i} + A_5 k + A_6 \log(k) + A_7 \log(1-k), \qquad (9.20)$$

where $k(S[t]) = K + S[t](1 - 2K)$, and $K = 0.15$. The voltage and current profiles are presented in Figures 9.2 and 9.3. As shown in these figures, the charging voltage reaches $v_m = 4.2V$, and remains constant for the rest of the duration. Since $\theta = 0.5$ puts more weight on the charging duration, it reaches the maximum allowed voltage quickly. Similarly, the initial charging current is higher for the $\theta = 0.5$, and it remains constant until $t_k = 2300$ seconds. During the rest of the charging, the

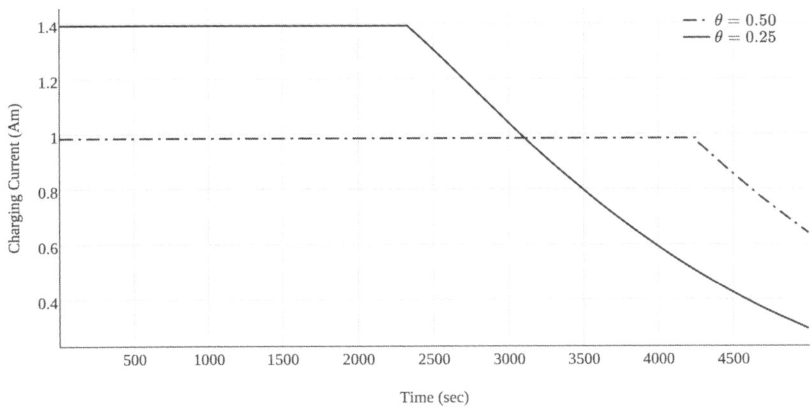

Figure 9.3 Optimal current profile.

charging current decreases and approaches zero. The final result of this section, depicted in Figure 9.4, shows the corresponding SoC time curve. It is observed that as SoC level increases, charging time also increases. For instance, it takes 20 minutes to charge the battery from 20% to 40%, while it takes around 85 minutes to charge the battery from 80% to 100%. This is due to the decrease in charging current, and subsequently, charging power.

9.3 Charging Control at a Fast-Charging Station

The previous section was focused on analyzing only the well-known CCCV charging strategy for Li-ion batteries. In reality, however, PEV owners might decide to take into account additional performance measures such as the charging price, the battery degradation, service waiting time, and even service outage rates. Therefore, in this section, we consider a small-scale fast public charging station and aim to maximize its operational profit. The main idea is that instead of allowing customers to fully charge their batteries, the station limits the maximum SoC level so that the service time for an average vehicle is reduced. As a result, more customers will be

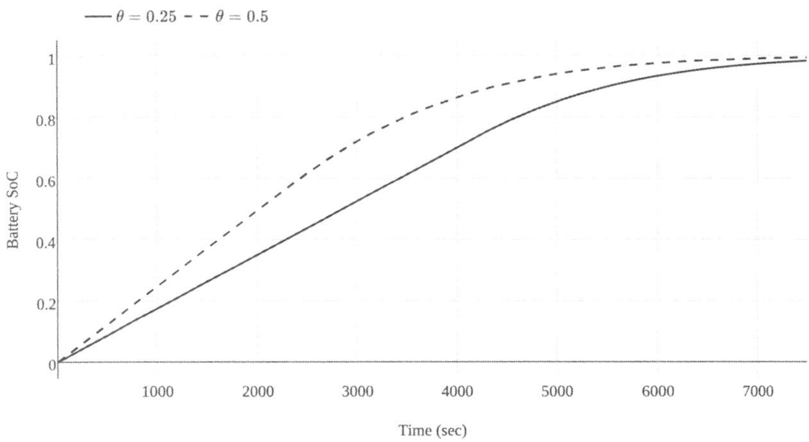

Figure 9.4 Battery state of charge curve.

served with the same amount of resources and customers will receive an acceptable level of battery charge (e.g., 90%).

In the previous chapter, the charging station was assumed to impose no waiting time. However, in a more realistic setting, demand might exceed resources, which leads to the imposition of waiting time. Hence, we model the charging facility as a $M/G/C/K$ queue, in which there are C charging poles and the station has a total PEV capacity of K customers, meaning that there are $r = K - C \geq 0$ waiting *spaces*, which is typically around $3-4$. It is noteworthy that if there are more than K customers in the station, the next service request will not be accommodated, and as a result it will be blocked. Hence, both the waiting time and the blocking probability serve as natural and relevant performance metrics.

Moreover, the customers arrive at the station according to a Poisson process with rate λ and general charging time distribution [12, 13], which is a function of the target SoC levels and the charging power. In the previous section, we showed that the charging current is constant until a threshold is reached, and then for the rest of the charging duration the voltage is kept constant while the charging current decreases. In this section, we translate our findings into charging power that is measured in kilowatts and adopt the charging curve specified by ABB fast chargers presented in [14]. Even though the product of voltage and the current is not constant,

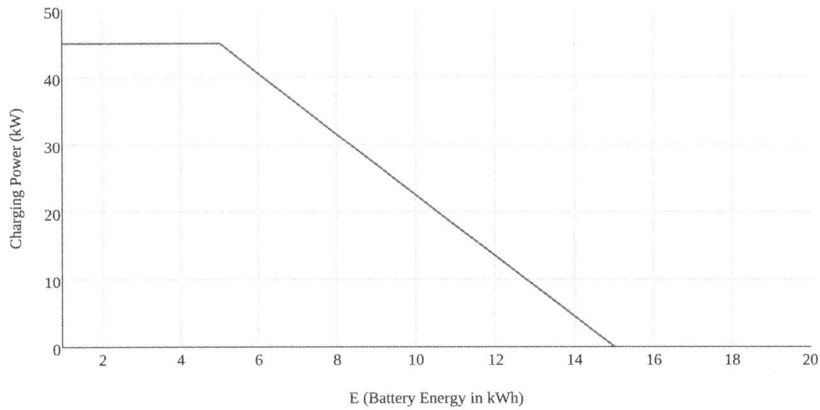

Figure 9.5 Battery charging power [13].

often in the the literature the charging power is linearized, according to which the battery is charged with a fixed power until it reaches a threshold E_c, then the power reduces as the SoC increases. For instance, the studies presented in [14] and [12] consider realistic scenarios, and show that the charging power as a function of stored energy ($P(E)$) is well-represented with the following piecewise linear relationship.

$$P(E) = \begin{cases} P_{\max} & E < E_c \\ \alpha_1 - \alpha_2 \cdot E & \text{otherwise} \end{cases}, \tag{9.21}$$

where $P_{\max}$ is the maximum allowed charging power, which is typically determined by the charger or the battery type. E_c is the stored energy when the battery OCV reaches the maximum voltage v_m. Moreover, α_1 and α_2 are constant parameters specified by the battery type. A typical charging curve is presented in Figure 9.5.

As discussed earlier, charging durations are linked to the initial state of charge (E_i), the charging power $P(E)$, and the target SoC level E_0 set by the station operator. It is further assumed that the initial SoC distribution can be obtained from the historical customer data, which is known to the station operator. Then, the charging duration for an arbitrary initial SoC level E_i to an objective SoC level E_0 can be calculated by $t = t_1 + t_2$, where

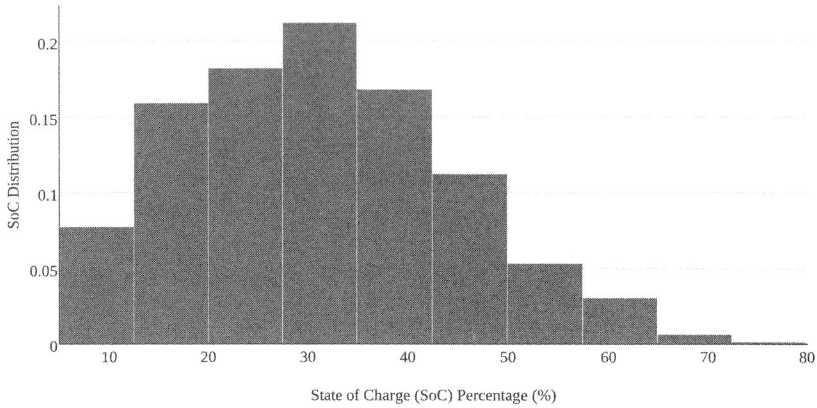

Figure 9.6 Initial state of charge distribution of arriving customers.

$$t_1 = \begin{cases} \frac{E_c - E_i}{P_{\max}} + \frac{1}{n_1} \log\left(\frac{\alpha_1 - \alpha_2 \cdot E_c}{\alpha_1 - \alpha_1 \cdot E_o}\right), & E_i \leq E_c \\ 0, & \text{otherwise} \end{cases}, \tag{9.22}$$

and

$$t_2 = \begin{cases} \frac{1}{\alpha_2} \log\left(\frac{\alpha_1 - \alpha_2 \cdot E_i}{\alpha_1 - \alpha_2 \cdot E_0}\right), & E_i \geq E_c \\ 0, & \text{otherwise} \end{cases}, \tag{9.23}$$

Note that t_1 represents the duration that it takes the battery to be charged with $P_{\max}$ until the SoC reaches E_c, while t_2 provides the service duration for the rest of the charging until the SoC reaches E_0.

For illustration purposes let us assume that all arriving customers have a PEV package of 16 kWh. The initial SoC distribution of vehicles follows a Gaussian distribution with a mean of 30 and standard deviation 15. To avoid negative values, the distribution is truncated to $[5, 80]$. In order to show the effects of the target SoC (E_0), three different target levels, $E_0 = 85\%, 90\%$, and 95% are considered. As discussed earlier, the charging duration increases exponentially as the E_0 gets close to 100%. Therefore, limiting the E_0 reduces the overall service duration

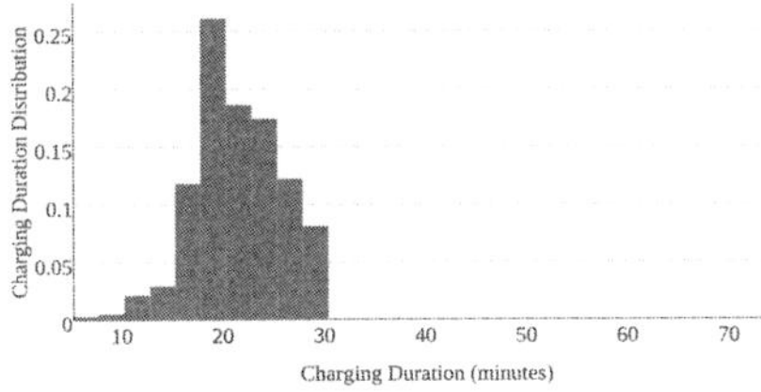

(a) SoC target 80%.

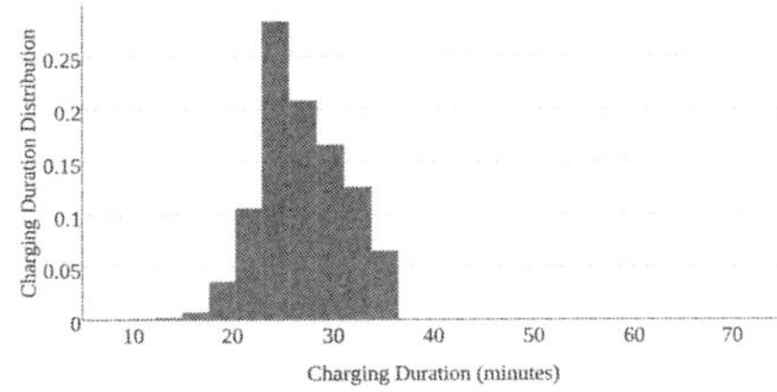

(b) SoC target 85%.

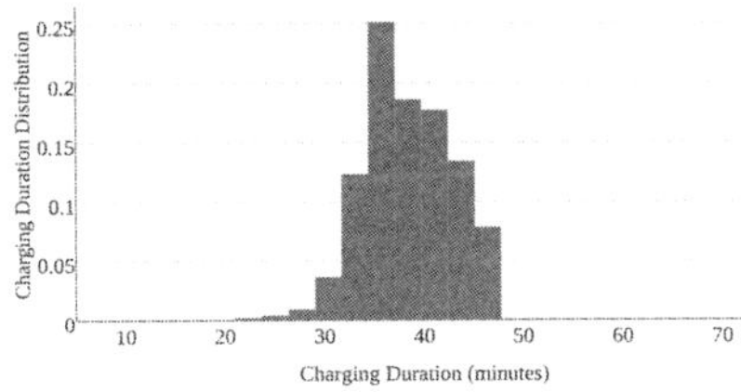

(c) SoC target 90%

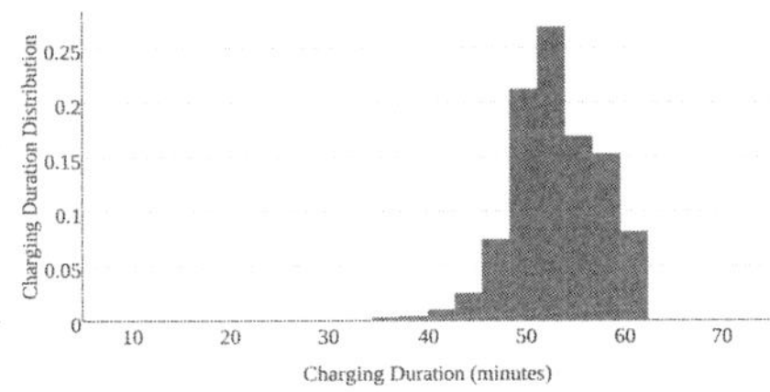

(d) SoC target 95%.

Figure 9.7 Charging duration of vehicles (16 kWh) with SoC distribution given in Figure 9.6.

and allows the station to charge more vehicles. Results presented in Figure 9.7 show that increasing the target state of charge from 85% to 95% almost doubles the charging duration. Moreover, for a given initial SoC level E_i and the average charging duration $\mathbb{E}(t)$, the average charging power, $\mathbb{E}(P_{\text{ch}})$ can be written as

$$\mathbb{E}(P_{\text{ch}}) = \frac{E_0 - E_i}{\mathbb{E}(t)} . \tag{9.24}$$

Notice that the average charging power is a function of the target SoC level, hence it plays a key role in a network optimization framework that will be discussed in the following sections.

9.3.1 Charging Station Queuing Analysis

Built on the analysis for charging durations and the effects of limiting SoC levels, next we assess the effects other detailed station performance metrics; namely, the waiting times and the blocking probability, based on the proposed $M/G/C/K$ queueing model. In the existing literature, approximation methods are proposed to determine the waiting times and the probability of blocking [12, 13, 15, 16]. The rationale underlying the approximation is that the closed-form expressions for $M/M/C$ and $M/D/C$ systems are well-established (infinite waiting space systems, former with exponential customers arrivals and the latter with deterministic arrivals). Hence, using the existing closed-form expressions of these two models, the probability of customer distribution in a $M/G/C$ queue can be obtained readily. In the next step, the blocking probability is computed by setting $K = C + r$ as the waiting space of the station. By defining $\rho = \frac{\lambda}{C\mu}$ as the traffic intensity of the station, the probability of having j vehicles in the station can be represented by [15, 16],

$$\mathbb{P}_j = \begin{cases} \frac{(C\rho)^j}{j!}\mathbb{P}_0 & j \in \{0, \cdots, c-1\}, \\ \frac{(C\rho)^C}{C!}\frac{1-\zeta}{1-\rho}\zeta^{j-S}\mathbb{P}_0 & j \in \{c, \cdots, K-1\}, \\ \frac{(C\rho)^C}{C!}\zeta^{K-C}\mathbb{P}_0 & j = K, \end{cases} \tag{9.25}$$

where

$$\mathbb{P}_0 \triangleq \left[\sum_{j=0}^{C-1}\frac{(C\rho)^j}{j!} + \frac{(C\rho)^C}{C!}\frac{1-\rho\zeta^{K-C}}{1-\rho}\right]^{-1}, \tag{9.26}$$

$$\zeta \triangleq \frac{\rho R_G}{1-\rho} + \rho R_G\,, \tag{9.27}$$

$$\text{and} \quad R_G \triangleq \frac{\mathbb{E}_W\,(M/G/C)}{\mathbb{E}_W\,(M/M/C)}, \tag{9.28}$$

where $\mathbb{E}_W(M/M/C)$ and $\mathbb{E}_W(M/D/C)$ are the expected waiting time of $M/M/C$ and $M/D/C$ systems, respectively. It is noteworthy that $\mathbb{E}_W(M/G/C)$ can be approximated according to the following closed-form expression:

$$\mathbb{E}_W\,(M/G/C) \approx \frac{1 + c_d^2}{\frac{2c_d^2}{\mathbb{E}_W(M/M/C)} + \frac{1-c_d^2}{\mathbb{E}_W(M/D/C)}}, \tag{9.29}$$

where c_d^2 is the variance of charging duration distribution. Therefore, R_G in (9.28) can be expressed as

$$R_G = \frac{\left(1 + c_d^2\right) R_D}{\left(2R_D - 1\right) c_d^2 + 1}, \tag{9.30}$$

where $R_D = \frac{\mathbb{E}_W(M/G/C)}{\mathbb{E}_W(M/D/C)}$ is the ratio of waiting times of $M/G/C$ and $M/D/C$ systems, which can be explicitly expanded as [15, 16]

$$R_D = \frac{1}{2}\left[1 + F(\theta)\, g(\rho)\left(1 - \exp\left\{-\frac{\theta}{F(\theta)\, g(\rho)}\right\}\right)\right], \tag{9.31}$$

where

$$\theta \triangleq \frac{C-1}{C+1}, C \geq 1, \tag{9.32}$$

$$F(\theta) \triangleq \frac{\theta}{8(1+\theta)}\left(\sqrt{\frac{9+\theta}{1-\theta}} - 2\right), \tag{9.33}$$

$$\text{and} \quad g(\rho) \triangleq \frac{1-\rho}{\rho}. \tag{9.34}$$

The final step in the analysis is to apply Little's law to compute the expected waiting time in the queue. According to Little's law $\mathbb{E}(L_q) = \lambda \cdot \mathbb{E}(t_w)$; that is, the value of expected waiting time $\mathbb{E}(t_w)$ can be represented in terms of the the expected number of PEV in the queue $\mathbb{E}(L_q)$. By adopting the approach of [15], we have

$$\mathbb{E}(L_q) = \sum_{j=C}^{C+r} (j - C)\mathbb{P}_j \tag{9.35}$$

$$\approx \frac{(C\rho)^C}{C!} \frac{\zeta}{(1-\rho)(1-\zeta)} \{1 - \zeta^r - r(1-\zeta)\rho\zeta^{r-1}\}\mathbb{P}_0. \tag{9.36}$$

Therefore the expected waiting time is

$$\mathbb{E}(t_w) = \frac{1}{\lambda}\frac{(C\rho)^C}{C!} \frac{\zeta}{(1-\rho)(1-\zeta)} \{1 - \zeta^r - r(1-\zeta)\rho\zeta^{r-1}\}\mathbb{P}_0. \tag{9.37}$$

9.3.2 Station Revenue Model

In this section we present an optimization framework to maximize the revenue obtained by serving vehicles. The primary goal is to find an optimum target SoC level E_0 so that the maximum number of vehicles can be served with the same amount of station resources. Each customer successfully entering the station gains a reward of R units, which is a function of reached SoC level according to

$$R = m \cdot \mathrm{SoC_r} + n \,, \tag{9.38}$$

where m and n the constant positive parameters. Furthermore, each customer has to pay a charging fee p. On the other hand, customers experience disutility due to waiting for service as well as battery degradation caused by high charging power. To formalize the analysis, we adopt the battery degradation model presented in Chapter 4 (Section 4.1.4). The expected battery degradation cost c_{batt} as a function of the charging power can be expressed as

$$\mathbb{E}(c_{\text{batt}}) = \beta_1 \cdot (c_{\text{batt}}))^2 + \beta_2 \cdot \mathbb{E}(P_{\text{ch}}) + \beta_3 \,. \tag{9.39}$$

Then, the aggregate degradation incurred during the mean charging duration is $\mathbb{E}(c_{\text{batt}}) \cdot \mathbb{E}(t)$. The total cost for a customer can be expressed as

$$C_{\text{total}} = p + c_w \cdot \mathbb{E}(t) + \mathbb{E}(c_{\text{batt}}) \cdot \mathbb{E}(t) \,, \tag{9.40}$$

where c_w is the unit cost per waiting in the system. The station operator can increase the admission price to earn more system profit [12, 13], while the customers accept to join the system only if their reward R is larger than the total cost C_{total}; that is,

$$R \geq p + c_w \cdot \mathbb{E}(t) + \mathbb{E}(c_{\text{batt}}) \cdot \mathbb{E}(t). \tag{9.41}$$

In other words, the reward should be equal or larger than the combined costs associated with the admission fee, waiting for the service, and battery degradation. Hence, the maximum admission fee paid by a PEV is

$$p = R - c_w \cdot \mathbb{E}(t_w) - \mathbb{E}(c_{\text{batt}}) \cdot \mathbb{E}(t) \,. \tag{9.42}$$

From the station operator standpoint, the goal is to maximize the station revenue. In unit time λ vehicles arrive at the station, $(1 - \mathbb{P}_K)$ fraction of them are admitted for service by the system, and each vehicle pays p units of the fee. Based on this model, station revenue becomes

$$R_{\text{station}} = \lambda \cdot (1 - \mathbb{P}_K) \cdot p \,. \tag{9.43}$$

By plugging (9.42) into (9.43), the revenue optimization problem can be expressed as maximizing the following revenue

$$R_{\text{station}} = \lambda \cdot (1 - \mathbb{P}_K) \cdot (R - c_w \cdot \mathbb{E}(t) - \mathbb{E}(c_{\text{batt}}) \cdot \mathbb{E}(t)) \,, \tag{9.44}$$

where based on the discussions in the previous section, the blocking probability is

$$\mathbb{P}_K = \frac{(C\rho)^C}{C!} \varsigma^{K-C} \mathbb{P}_0 \,. \tag{9.45}$$

9.3.3 Numerical Evaluation

We present a case study to compare the system performance with that of the baseline scenario in which each vehicle is charged up to 99%. The simulation parameters are selected as follows. The station employs $C = 7$ charging poles and $r = 3$ waiting spaces. The initial state of charge distribution is $\mathcal{N}(30, 15)$ truncated to $[5, \text{SoC}(E_0 - 10\%]$. The optimal state of charge level for an objective E_0 can be found in a finite number of iterations by discretizing the SoC levels at 1% resolution. The target $\text{SoC}(E_0$ is assumed to lie in the range of 60%-95%. Moreover, the charging power parameters are set as $P_{\max} = 45kW$, $\alpha_1 = 67.5$, and $\alpha_1 = 4.5$. The battery degradation parameters are taken from Chapter 4 and set as $\beta_1 = 0.004$, $\beta_2 = 0.0075$, and $\beta_3 = 0.003$. The cost of waiting one hour is \$20, and the revenue parameters are $m = \$9$ and \$2.

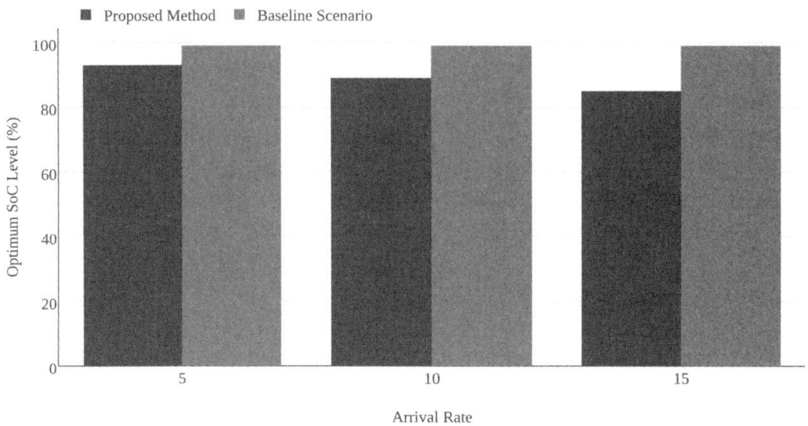

Figure 9.8 Optimum target state of charge levels.

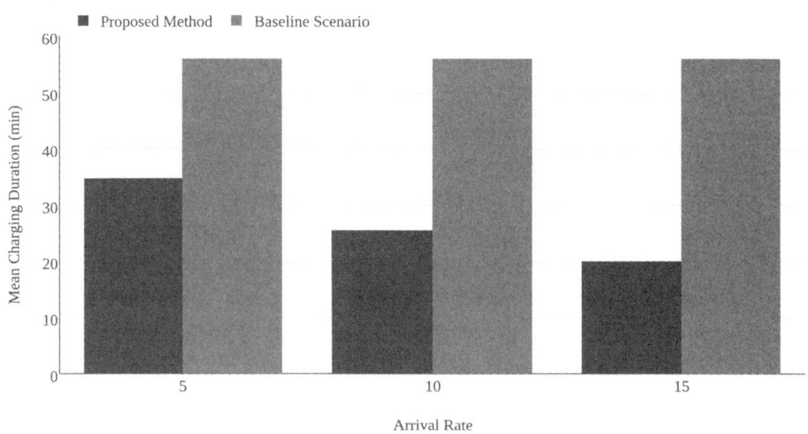

Figure 9.9 Mean charging duration for customers.

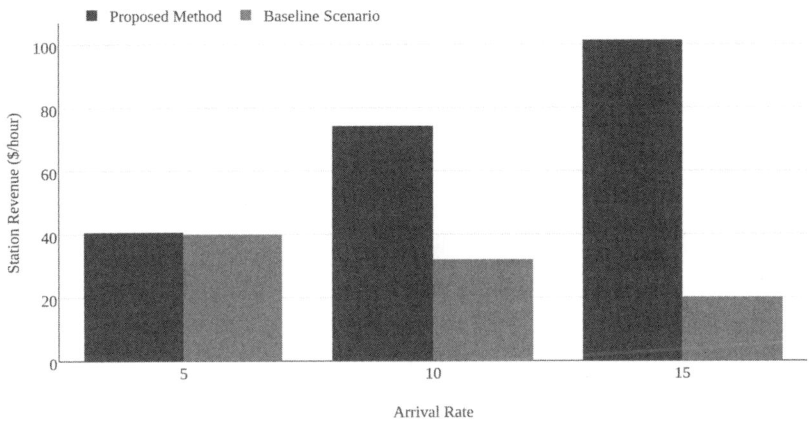

Figure 9.10 Comparison of station revenues.

We evaluate system performance for the three different arrival rates $\lambda = 5$, 10, and 15 per hour. In Figure 9.8, we show the optimal SoC levels that maximize the station revenue. The target SoC reduces as the arrival rate increases because the station operator needs to reduce the expected waiting time and the blocking probability to accommodate additional demand. Similarly, the expected charging durations decrease, but when compared to the baseline scenario, the proposed model outperforms. This is illustrated in Figure 9.9.

Station revenues are also compared and presented in Figure 9.10. Since the proposed model adjusts the SoC levels according to the traffic demand, station revenue increases when compared with the baseline case. On the other hand, under light traffic regimes, the two systems perform similarly. This is due to the fact that station resources can serve all arriving customers with a negligible blocking probability.

9.4 Chapter Summary

In this chapter, we provided two mathematical frameworks on PEV charging rates. In the first one, we considered charging of a single PEV battery. The objective function included two competing components: minimizing energy losses and shortening the charging duration. In the second framework, charging power should be increased, which leads to increasing the losses, which are proportional to the current passing through the resistor. The optimal charging rates determine the constant current constant voltage charging regime. Furthermore, later in the chapter, we considered a fast-charging station with waiting space. The goal was to maximize station revenue by optimally computing the target state of charge levels for the customers. From a customer standpoint, the goal was to fill up the battery by also considering the cost associated with battery degradation, waiting time, and blocking probability. The simulation results showed that limiting SoC level significantly improves system performance and maximizes station revenue.

References

[1] Rotering, N., & Ilic, M., "Optimal Charge Control of Plug-in Hybrid Electric Vehicles in Deregulated Electricity Markets," *IEEE Transactions on Power Systems*, Vol. 26, No. 3, 2010, pp. 1021-1029.

[2] Fan, Z., "A Distributed Demand Response Algorithm and Its Application to PHEV Charging in Smart Grids," *IEEE Transactions on Smart Grid*, Vol. 3, No. 3, 2012, pp. 1280-1290.

[3] Acha, S., Green, T. C., & Shah, N. (2010, April), "Effects of Optimised Plug-In Hybrid Vehicle Charging Strategies on Electric Distribution Network Losses," *Proc. IEEE PES Transmission and Distribution Conference*, New Orleans, LA, Apr. 2010, pp. 1-6.

[4] Sortomme, E., et al., "Coordinated Charging of Plug-in Hybrid Electric Vehicles to Minimize Distribution System Losses." *IEEE Transactions on Smart Grid*, Vol. 2, No. 1, 2011, pp. 198-205.

[5] Clement, K., Haesen, E., & Driesen, J., "Coordinated Charging of Multiple Plug-in Hybrid Electric Vehicles in Residential Distribution Grids," *Proc. IEEE/PES Power Systems Conference and Exposition*, Seattle, WA, Mar. 2009, pp. 1-7.

[6] Young, K., et al., "Electric Vehicle Battery Technologies". in *Electric Vehicle Integration into Modern Power Networks*, Rodrigo Garcia-Valle and Joao Pecas Lopes (eds.), New York: Springer New York, 2013, pp. 15-56.

[7] Abdollahi, A., et al., "Optimal Battery Charging, Part I: Minimizing Time-to-Charge, Energy Loss, and Temperature Rise for Ocv-Resistance Battery Model," *Journal of Power Sources*, Vol. 303, 2016, pp. 388-398.

[8] Verriest, E. I., & Lewis, F. L., "On The Linear Quadratic Minimum-Time Problem," *IEEE Transactions on Automatic Control*, Vol. 36, No. 7, 1991, pp. 859-863.

[9] El Alami, N., Ouansafi, A., & Znaidi, N., "On The Discrete Linear Quadratic Minimum-Time Problem", *Journal of the Franklin Institute*, Vol. 335, No. 3, 1998, pp. 525-532.

[10] Bertsekas, D. P., *Nonlinear Programming*, Belmont, MA: Athena Scientific, 1999.

[11] Balasingam, B., et al., "A Robust Approach to Battery Fuel Gauging, Part I: Real Time Model Identification," *Journal of Power Sources*, Vol. 272, 2014, pp.1142-1153.

[12] Fan, P., Sainbayar, B., & Ren, S., "Operation Analysis of Fast Charging Stations with Energy Demand Control Of Electric Vehicles," *IEEE Transactions on Smart Grid*, Vol. 6, No. 4, 2015, pp. 1819-1826.

[13] Kong, C., Bayram, I. S., & Devetsikiotis, M. (2015). "Revenue Optimization Frameworks for Multi-Class PEV Charging Stations", *IEEE Access*, Vol. 3, 2015, pp. 2140-2150.

[14] Andersson, D., & Carlsson, D., "Measurement of ABB's Prototype Fast Charging Station for Electric Vehicles", master's thesis in electric power engineering, Chalmers University of Technology, 2012

[15] Kimura, T., "Approximations for Multi-Server Queues: System Interpolations," *Queueing Systems*, Vol. 17, No. 3-4, 1994, pp. 347-382.

[16] Kimura, T., "A Transform-Free Approximation for the Finite Capacity M/G/s Queue," *Operations Research*, Vol. 44, No. 6, 1996, pp. 984-988.

Chapter 10

Charging Facility Location Design

10.1 Introduction

Designating the locations of the charging facilities is instrumental to enhancing the mainstream acceptance of PEVs. Specifically, customers' reliance on PEVs is strongly related to having well-planned charging infrastructure that can accommodate a desirable number of trips. Designing the locations essentially involves two entangled problems that can affect the mainstream adoption of PEVs. On the one hand, PEV adoption is ultimately affected by convenient access to charging stations, which necessitates having a larger number of charging options. On the other hand, investors are less likely to invest in building charging infrastructures unless adoption reaches a certain level. Besides these two key issues, designing the locations of the charging stations also affect the convenience in accessing PEV charging facilities, which in turn will impact the demand for petroleum, electricity consumption, and power grid operations. Hence, the proper expansion of PEV infrastructure impacts carbon emissions and reliance on fossil fuels for transportation.

Motivated by the need for planning charging station locations, location theory, which in general addresses questions pertinent to the locations of economic activities, provides a relevant framework for designing charging station locations. Specifically, location theory identifies the locations of facilities in order to optimize quantifiable desirable objective functions that are linked to the location of the facilities. Even though location theory has inspired a rich body of literature [1–10], the body of research on designing the locations of the PEV charging facilities is sparse and requires further investigation [11–14]. The main reason is the fact that the PEV penetration rates have not reached expected levels yet. Therefore, the most important challenges associated with designing the PEV station locations require further scrutiny and investigation. In this chapter, we provide an overview of the facility location problems. Moreover, we present two case studies, where the first one focuses on the locations of slow-charging stations in parking lots, while the second one considers a highway network and analyzes fast-charging station settings.

The history of facility location theory dates back to 1909 with the pioneering studies of Alfred Weber, in which he considered the siting problem of a single warehouse in order to minimize the total distance between the warehouse and the customers. Later, Hotelling studied facility location of two facilities on a line [5]. In this simple model, customers are uniformly distributed along the line model and they prefer to use the nearest facility. In this setting, the optimal locations for both facilities are found to be the middle points. The interest in facility location problems was renewed in the mid-1960s with the expansion of the telecommunication networks, in which the goal was determining the optimal locations of switching facilities with the purpose of minimizing the cost of the required infrastructure. The interest in the analytical aspects of the location problem grew extensively over the subsequent years, during which researchers focused on both formulation and algorithm development for a diverse set of problems arising in a number of domains in the private and public sectors. Some of the most popular private sector domains include designing the locations of banks/ATMs, industrial plants, and retail facilities. In the public sector, on the other hand, they are primarily focused on designing locations of medical emergency centers, fire stations, schools, and hospitals [2–5]. The distinctive aspect of the public and private sectors in facility location problems is that the private sector problems aim to find sites in order to minimize monetary costs, while in the public sector, the primary goal is to optimize access time by the population. Hence, the objectives of the private and the public sectors may not necessarily always align perfectly. The same public and private dichotomy also appears in the context of designing charging station locations. The currently existing charging stations are mostly operated and owned by the private sector, while being at the same time incentivized by the government. Hence, location problems may need to consider competing, and even perhaps opposing, objectives, where one tends to reduce the siting cost and another one to improve accessibility.

The ramifications of deploying charging stations are long-lasting and impact various operational decisions. For the case of fast-charging facilities, the high costs associated with land property acquisition, electrical equipments, energy storage systems, and facility construction make such siting projects long-term investments. Moreover, the current state of the industry, such as vehicle distribution, driving routes between the points of interest, and population concentrations, might change in the future. Therefore, finding robust station locations can be complex, which needs to incorporate costs associated with uncertain future events. In contrast, siting slow-charging stations, namely level 1 and 2 stations, will mostly take place in the currently existing public spaces such as parking lots, shopping malls, and university campuses, in which case the location problem is less complicated.

Designing charging station locations involves analyzing and balancing four principal components; (1) concentration of PEV users that exist in a certain geographical area; (2) the charging facilities to be installed; (3) a space in which facilities and customers are stationed; and (4) a metric that measures the distances or driving time between PEVs and the stations. Such analyses require adopting proper models for station locations and aim to answer the following questions:

(a) How many charging facilities should be sited to guarantee a pre-specified quality of service goals?

(b) Where should each facility be located in a given region?

(c) What should be the capacity of each station?

(d) What type of chargers should be deployed?

(e) Should the charging stations be located as clusters or sets of interconnected regions?

Furthermore, additional challenges might emerge due to impending developments in the industry, such as capacity planning based on the availability of renewable and storage units at each station. The answers to these research challenges depend on the context in which facility locations is formulated and solved. For instance, in major metropolitan cities, city planners may favor installing faster-charging facilities as drivers may not have access to parking lots for extended periods of time. On the other hand, the physical land in small and college towns makes slow-charging facilities more appropriate options, as the driving ranges of the daily commutes are short.

Before formalizing a framework for addressing these questions, it is noteworthy to mention that there also exists an inherent trade-off between service and cost in facility location problems. If excessive number of charging stations are deployed, the aforementioned costs are likely to exceed the desired benefits, while installing too few facilities degrades the quality of the charging service. As mentioned earlier, poor location decisions will also impact PEV penetration levels. Hence charging station location problems have a critical role in the realization of sustainable transportation systems.

10.1.1 Sample Objectives

In this section, we present several examples that motivate incorporating different performance guarantees into PEV charging station problems. The first example considers the case in which a system operator (public or private) aims to provide a charging service in a given geographical area (e.g., an entire city or specific regions) so that any vehicle can reach a station by driving under X miles. Therefore, the first objective is to *minimize the number of charging stations required so that all customers can reach a charging facility within the required driving distance.* Suppose that a charging station operator announces that any customers can access at least one station within $X = 10$ miles of any desired location. Achieving this objective, which is also studied under the *set covering problem*, will substantially boost customer confidence in beating range anxiety.

The approaches that address incorporating such maximum driving range for accessing a charging station can potentially lead to underutilization of charging stations that will lead to costly operation. The main reason is that the system operator may have to site stations to serve low-density areas, which might reduce profit margins. Suppose that deploying N stations satisfies the X miles requirement. Under this setting, the penalty for removing one station is that the network will be not serving a part of the demand. However, this penalty is significantly less than $1/N$ [6]. Therefore, an alternative objective can be to *maximize the amount of customer demand that can be served using a given number of stations.* This problem is also known as the *maximum covering problem*, an example of which is depicted in Figure 10.1. In this example, eight facilities support the entire customer demand. However, if the station operator sacrifices to reject a small percentage of the customers, the deployment cost reduces by about $1/8$. The third model is motivated by accounting for the cost-coverage trade-off. Assuming that the system operator has a fixed budget and can afford to deploy P stations, the goal is to *minimize the maximum driving distance with P stations.* Such models are referred to as *P-center problems* and solve the siting problem for a scenario in which customers

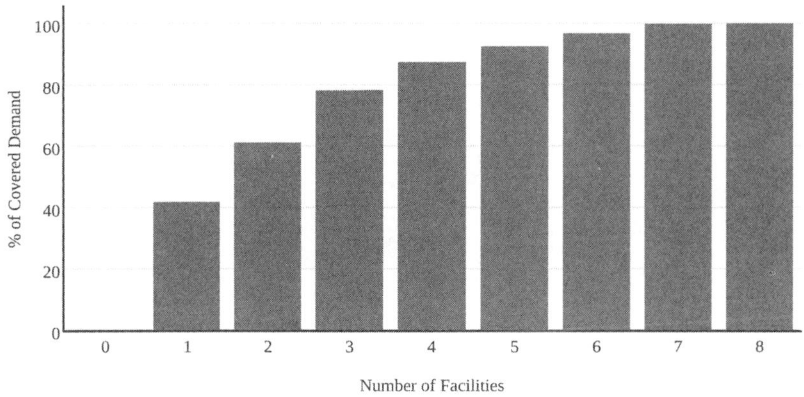

Figure 10.1 Facility coverage trade-off for basic coverage problem. Data set is presented in [6].

require the highest possible demand. Therefore, this guarantees an upper bound for system performance, which can be, for instance, the driving distance.

The aforementioned three models all focus on ensuring QoS guarantees for the worst-case scenarios. On the other hand, an alternative approach is to focus on the statistical measures and consider *average* driving distance. Under such measures, stations are stationed in order to *minimize the average driving distance for a given number of charging stations (P)*. Finally, the "average" can be replaced by "median," leading to the P-median problem. An overview of discrete network models is presented in Figure 10.2.

The objectives given above can be considered as the basis for addressing charging station siting problems, but the real-world implementations may require further detailed modeling. For instance, the described models do not consider the random nature of customers' behavior. Therefore, instead of average driving distance, one might be interested in the distribution of driving distances. Similarly, hard deterministic requirements can be replaced by probabilistic guarantees that customers will be within X miles driving distance from the charging stations. Moreover, there could be multiple classes of customers demanding different charging

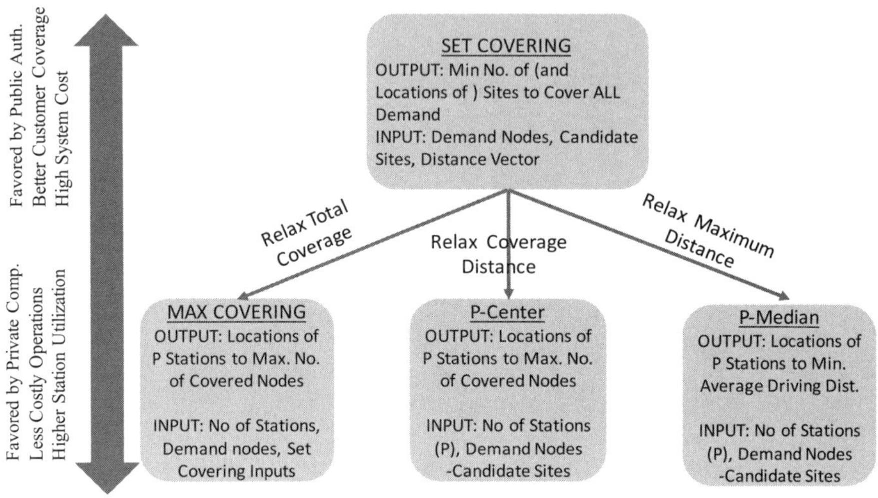

Figure 10.2 An overview of discrete location models.

rates or charger types. Therefore, system planners may need to take into account more comprehensive and multi-objective decision-making procedures.

10.2 Classification of Facility Siting Problems

Facility location models can be classified in a number of ways and this section aims to provide an overview of a commonly accepted taxonomy.

10.2.1 Continuous and Discrete Network Models

Continuous location models (also referred as planar location models) assume that facilities (PEV stations) and customers can be located anywhere on a plane, and subsequently, the solution space is continuous. Since customers of fast-charging stations are likely to be mobile, the facility location problem can be properly modeled by such continuous planar structures. A representative study on such models is presented in the next section in which vehicles are assumed to travel on continuous highways. The continuous location problems usually employ d-dimensional real spaces, and distances are calculated according to Minkowski distances with the parameter p. For instance, the ℓ_p distance between two distinct points (x_i, y_i) and (x_j, y_j) with $i \neq j$ is defined as

$$d_{ij}^p = [|x_i - x_j|^p + |y_i - y_j|^p]^{1/p} . \tag{10.1}$$

The vast majority of such continuous models consider three different choices for p:

(a) $p = 1$: corresponds to the ℓ_1 norm and represents the Manhattan, rectangular, or right-angle distance metric; that is,

$$d_{ij} = |x_i - x_j| + |y_i - y_j| .$$

(b) $p = 2$: corresponds to the ℓ_2, Euclidean, or straight-line distance metric:

$$d_{ij}^2 = \sqrt{|x_i - x_j|^2 + |y_i - y_j|^2} .$$

(c) $p = \infty$: corresponds to the ℓ_∞ or Chebyshev metric:

$$d_{ij}^\infty = \max\{|x_i - x_j|, |y_i - y_j|\} .$$

Recall that the classical Weber problem aims to determine the set of coordinates (x_s, y_s) for a single facility in order to minimize the weighted sum of distances of the customers to the facility. By defining $d_n(x_s, y_s)$ as the Euclidean distance between customer n located at (x_n, y_n) from the facility located at (x_s, y_s), $n \in N$, the facility location can be found as the solution to the following optimization problem.

$$\underset{(x_s, y_s)}{\text{minimize}} \sum_{n \in N} w_n d_n .(x_s, y_s) \tag{10.2}$$

This problem can be solved efficiently via an iterative procedure. On the other hand, if we include the additional constraint that the number of stations P must satisfy $P < |N|$, the multisource Weber problem can be formulated as

$$\begin{aligned} \text{minimize} \quad & \sum_{n \in N} \sum_{p=1}^{P} w_n d_n(x_p, y_p) a_{np} \\ \text{s.t.} \quad & \sum_{p=1}^{P} a_{np} = 1, \ \forall n \in \{1, \dots, N\} \\ & a_{np} \in \{0, 1\}, \ \forall n \in N, p \in \{1, \dots, P\} \end{aligned} , \tag{10.3}$$

where a_{np} is an indicator and is set to 1 if the demand n is assigned to facility p, and it is zero otherwise.

In discrete models, which are also sometimes referred to as network location models, stations and customers are assumed to be only on a network composed of discrete nodes and links. Such models are appropriate for siting slow-charging stations (e.g., level 1 and 2), as they will mostly be placed at already existing parking lots in which the demand is stationary during the charging service. Unlike the distance metrics mentioned above, discrete models use shortest path algorithms as the distance metric. As mentioned in the previous section, there can be a variety of objectives such as in the covering problems, median problems, and center problems.

Next, we provide an example formulation of the P-median problem. For this purpose, let N denote the set of nodes in the network, $\mathcal{S} \subset \mathcal{N}$ is the set of potential facilities. We also introduce a binary decision variable b_p, which is equal to one if site p is selected, and it is zero otherwise. Then the P-median problem can be formulated as

$$\begin{aligned} \text{minimize} \quad & \sum_{n\in N}\sum_{p\in S} w_n d_n a_{np} \\ \text{s.t.} \quad & \sum_{p\in S} a_{np} = 1, \ \forall n \in \mathcal{N} \\ & a_{np} - b_p \leq 0, \ \forall n \in \mathcal{N}, j \in \mathcal{S} \\ & \sum_{j\in S} b_p = P \\ & a_{np}, b_p \in \{0,1\}, \ \forall n \in \mathcal{N}\ , \ j \in \mathcal{S} \end{aligned} \quad , \tag{10.4}$$

Note that the first constraint guarantees that the demand is served, while the second inequality constraint couples the assignment and location decisions. The third constraint ensures that there are P stations to place. Similar to the P-center problem, the objective is to minimize the maximum distance r, hence the discrete optimization model can be written as

$$\begin{aligned} \text{minimize} \quad & r \\ \text{s.t.} \quad & r - \sum_{p\in S} w_n d_n a_{np} \geq 0 \ \forall n \in \mathcal{N} \\ & \sum_{p\in S} a_{np} = 1, \ \forall n \in \mathcal{N} \\ & a_{np} - b_p \leq 0, \ \forall n \in \mathcal{N}, j \in \mathcal{S} \\ & \sum_{j\in S} b_p = P \\ & a_{np}, b_p \in \{0,1\}, \ \forall n \in \mathcal{N}\ , \ j \in \mathcal{S} \end{aligned} \tag{10.5}$$

10.2.2 Static and Dynamic Problems

Charging station problems can also be classified based on the static or dynamic nature of the setting. The location problems given above are *static*, meaning that the system parameters do not vary over time. On the other hand, the input and output of *dynamic* models fluctuate over time. For instance, customer demands can understandably be different throughout the day and also can vary over different days of the week. Therefore, in addition to finding the optimal locations, dynamic models also answer the question of *when* to close an existing station or invest in new ones. In dynamic problems, the goal is to develop methods to find optimal operation range, rather than finding an optimal solution [8]. The dynamic nature of most charging station operations requires constant relocation of the facilities. In order to circumvent this shortcoming, a remedy is to create a cooperation among the stations by means of routing customers to the nearest available charging station.

10.2.3 Other Classification Methodologies

The charging station location problems can also be classified according to the following aspects.

- *Deterministic versus Stochastic Models:* The inputs of the charging station location problem could be deterministic (certain) or probabilistic (uncertain) [3]. For instance, if the station is located in a neighborhood with a fixed number of vehicles, then the system could be considered as deterministic. However, in many other cases stations have probabilistic inputs. For instance, consider stations in public areas such as shopping malls, hospitals, and airports, in which customer arrival patterns are highly uncertain. Therefore, problem formulations include expected values of the system parameters and the problems given above become equivalent to the *maximum expected to cover problem*, *α-reliable P-center problem*, and *maximum reliability location problem.*

- *Single versus Multiple Objectives:* The models described in the previous sections mostly contain single objectives, often pertinent to the driving distance. However, the siting problem may include multiple objectives. For instance, policy makers may require companies to serve a certain fraction of the population, which may conflict with the objective of minimizing the deployment cost.

Table 10.1
Key Statistics of Passenger Trips in Seattle [1]

(% of) Activity	Mean	Standard Deviation	Mean Parking Duration (min/trip)
Work	0.271	0.445	379.7
Personal business	0.179	0.384	46.8
Shopping	0.168	0.374	27.7
Pick-up/Drop-off	0.133	0.34	15.5
Dining	0.071	0.257	46.1
Recreation	0.057	0.232	103.5
Social	0.04	0.197	127.6
College	0.00763	0.087	222.5
School (K-12)	0.00687	0.083	338.8
Parking duration (min/trip)	142	199.5	N/A
Trip distance in miles	6.71	7.14	N/A

- *Single-level versus Tiered Models:* If there is no power flow and coordination among stations, the siting problem is considered to be a single-level problem. On the other hand, the charging service could also be provided by a tiered architecture and the coordination could be carried on via storage units. One such example is depicted in Figure 3.3 of Chapter 3.

- *Single-class versus Multiple Class:* The location models can also be divided according to the type of customers being considered. The previous chapters have presented several station models serving single or multiple classes of customers. In multiple classes the facility location problem aims to site different charging technologies concurrently to meet the design goals.

10.3 Siting Slow-charging Stations

In this section, we present a facility location problem aiming to locate chargers to parking stations [1]. The study is based on the survey data collected by Puget Sound Regional Council in 2006 and contains 30,000 personal trips in Seattle, WA. More specifically, the data includes 10,510 individuals with an average age of 41.9 years old, 78% of whom have a driver's license, 21% of the population are students, and the average number of licensed drivers at each house is 1.69. Other statistics are given in Table 10.1. For instance, 27% of the trips are for commuting to work, and vehicles remain stationary for an average duration of 6.3 hours. In such circumstances, workplace charging could become popular among PEV drivers. In fact, several companies are offering workplace charging as a benefit to their employees using PEVs, since charging at work is usually less expensive compared to charging at home due to lower industrial electricity tariffs [11]. Other activities, such as commuting to school and attending social events, also have strong potentials for the park-and-charge charging.

The goal of the study presented in [1] is to maximize the coverage of demand measured in parking minutes per mile square. Moreover, due to budget constraints, there exist P chargers to be placed at parking lots. The system operator aims to minimize the cost of moving vehicles from the intended parking space to the one that has a charger. Therefore, the system designer introduces another constraint on the maximum distance between stations with chargers and stations without chargers. The optimization problem is similar to the P-center location problem (with additional constraints), and it is formalized as follows [1].

$$\begin{array}{ll} \text{minimize} & \sum_i \sum_j c_{ij} y_{ij} \\ \text{s. t.} & \sum_j y_{ij} = d_i, \ \forall j \in J \\ & \sum_i y_{ij} = A x_j, \ \forall i \in I \\ & \sum x_j \leq P, \ \forall j \in J \\ & \sum_i \delta_{ij} x_j \leq 1, \ \forall i \in I \\ & y_{ij} \geq 0, \ \forall i \in I, \ \forall j \in J \\ & c_{ij} \leq C \\ & x_j \in \{0,1\}, \ \forall j \in J \end{array} \tag{10.6}$$

where the parameters are defined as follows.

Station ID	Demand (mins)	Demand Rank (out of 218)
70	8959	1
30	6449	2
90	5063	7
174	2466	17
101	2445	18
31	2311	19
13	2232	20
53	1481	40
39	771	74
56	749	76
93	678	77
199	603	85
209	593	87
37	552	93
180	421	106
152	376	109
162	341	113
65	118	152
19	69	167
153	0	207

Figure 10.3 Locating slow chargers at parking lots in Seattle, WA [11]. Traffic Analysis Zones (TAZ) are determined by Puget Sound Regional Council [15]. Traffic data is available at [16]. Stations are circled.

The case of 218 TAZ in downtown Seattle is considered for the location problem. The maximum walking distance, denoted by R, is set to $R = 2$ miles, the minimum distance between charging facilities, denoted by r is set as $r = 1$ mile, and the total number of chargers is set to $P = 20$.

As shown in Figure 10.3, stations are located primarily in zones with high demand, such as TAZs 70, 13, and 90. On the other hand, stations are also located in areas with low demand, but such stations are chosen due to their proximity to high-demand areas (e.g., 19, 65, and 152). The results indicate that 94.5% of the parking demand is met with a maximum walking distance of 1.53 miles (between

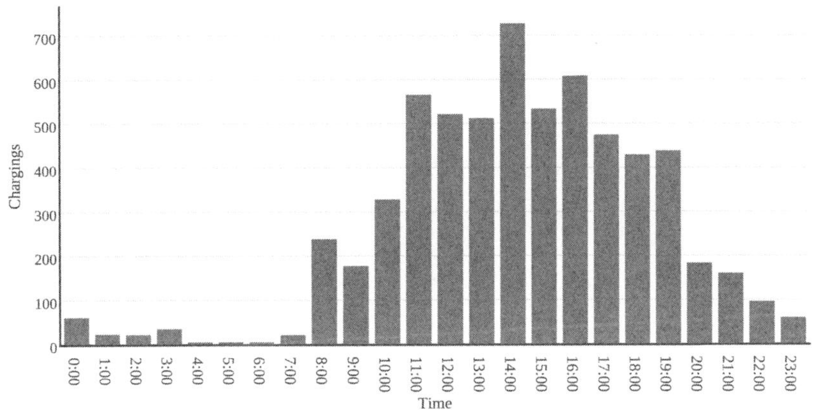

Figure 10.4 PEV Time of use demand on a university campus [17].

zones i and j). Note that placing $P = 20$ stations in zones with the highest demand covers 4.5% fewer PEVs.

Even though the case study given above provides rich insights into the deployment of charging stations at parking lots, it ignores the time dimension in analyzing customers demands. Furthermore, the above problem only considers determining the locations of the stations and it ignores how many chargers to deploy. The latter question is related to the *time of use* characteristics. In the simplest case, the system designer can decide on the number of chargers by analyzing the busiest hour. For instance, the study in [17] presents the PEV charging data in the parking lots of a major North American University for three years. Figure 10.4 shows that parking demand is high in the afternoon, while the morning demand is relatively low. Also, Figure 10.5 shows that demand during the weekdays is almost flat, while there is a significant decrease over the weekends.

10.4 Siting Fast-charging Stations

The previous section provided a detailed analysis of the facility siting problems in which the demand and the facilities are stationary. This approach is more suitable

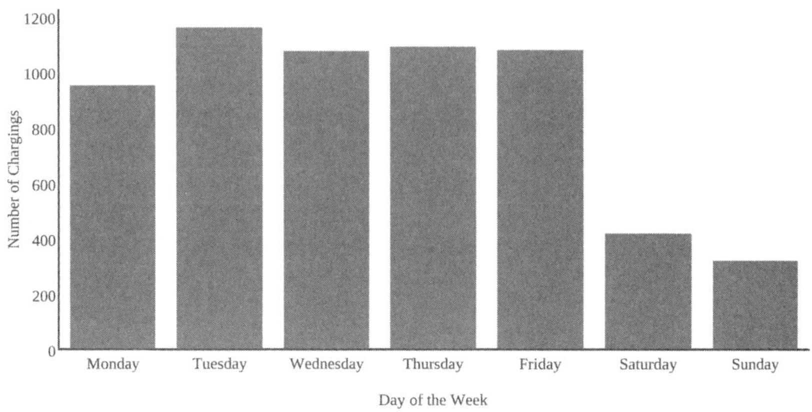

Figure 10.5 PEV daily demand on a university campus [17].

for level 1 and 2 charging stations as the vehicles are stationary for several hours in parking lots. The case for fast-charging stations, however, is fundamentally different due to the mobile nature of the vehicles. Drivers may not necessarily decide about their driving route only based on the locations of the charging facility and their relative distances to these stations. Instead, the deployment principle is often based on maximizing the *capture* of the vehicle *flow* between some origin and destinations. Flow-capturing location problems (FCLPs), first developed by Hodgson in 1990, provide mathematical frameworks for optimizing the amount of captured traffic flow [1]. An overview of siting problems is provided in Figure 10.6.

The FCLP has also attracted interest in deploying refueling stations for alternative charging resources, such as hydrogen and natural gas, to minimize the deployment cost of the fueling infrastructure network. One illustrative example is presented in Figure 10.7. In this example, two stations are located on a route of possible four origin-destination pairs, namely O-D1, O-D2, O-D3, and O-D4. The serving region of two stations are shaded and the traffic flows O-D2 and O-D3 have access to a charging station, while the remaining traffic flows will not be *captured* by the located facilities. The FCLP problem only considers the case when a vehicle is captured only one time during a trip. In reality, nevertheless, the case of electric transportation is more complex since driving ranges are limited and some vehicles

<u>SLOW CHARGING STATIONS</u>

- ✓ Level 1 and 2 chargers
- ✓ Long service durations: 3-8 hours
- ✓ Vehicles are parked and stationary
- ✓ Candidate locations are *discrete* parking lots
- ✓ Discrete network models are suitable for siting problems

<u>FAST CHARGING STATIONS</u>

- ✓ DC chargers
- ✓ Short service durations: 20-30 min
- ✓ Vehicles are mobile and traveling long distances
- ✓ Candidate locations could be built anywhere on the network
- ✓ Flow-capturing models are suitable for siting problems

Figure 10.6 Overview of PEV charging station location problems.

may need to recharge their batteries more than once. Hence, the goal is to place a number of charging stations that are adequate for completing a trip. This type of problems is referred to as the flow-refueling location model (FRLM).

To demonstrate the key points of the FRLM setting, consider the toy example given in Figure 10.8, in which a vehicle repeatedly travels 110 miles between the origin and destination points. It is assumed that the PEV is fully charged at the origin and we are interested in placing charging stations along the path. Clearly, the optimal locations of the charging stations depend on the all-electric range (AER) of the vehicle. When the vehicle has an AER of 220 miles (e.g. in Tesla models), then the station can be sited anywhere along the path, and irrespectively of the location of the charging destination, the vehicle can always reach its destination. In contrast, when the AER is 150 miles, it could happen that the capacity might not be sufficient to complete the trip. Therefore, locating a single node either at destination D or some middle point on B-D will be necessary. In the third case, consider a mainstream PEV with driving mileage of 85 miles. This case underscores an important point, which is that placing the charging stations exactly at the nodes O, A, B, or D is not optimal. To verify this by contradiction, let us assume that a station can be placed at only the nodal points. Based on this, in order to complete a round-trip drive, at least two stations are necessary, and they must be placed at nodes B and D. On the other hand, if the stations are allowed to be located in a middle point, one station placed between B and D will suffice to charge the vehicle at both directions. To be more specific, the station can be located anywhere that

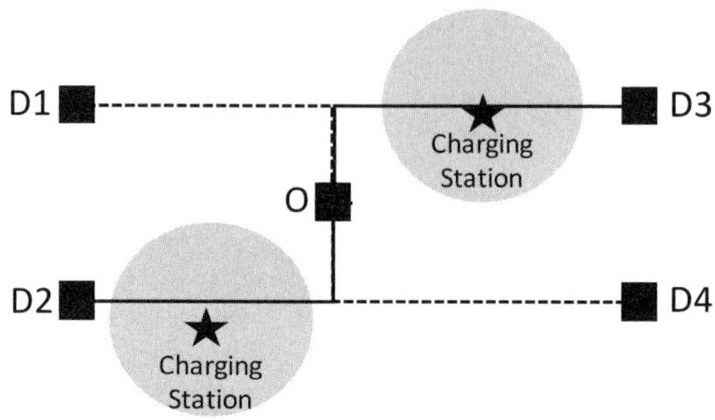

Figure 10.7 Overview of flow-capturing facility location problem.

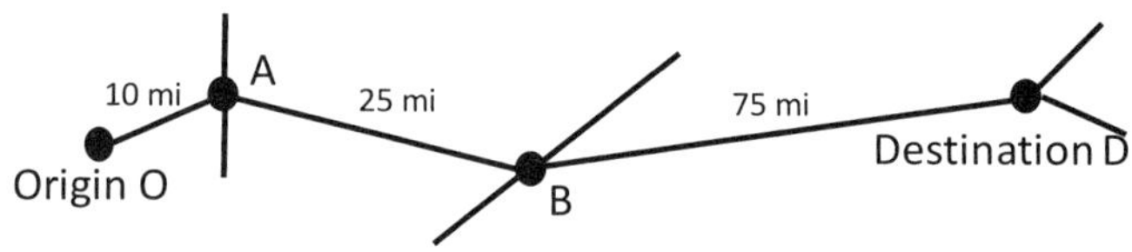

Figure 10.8 Toy example for an origin-destination traffic flow.

has a distance of 25 to 42.5 miles to D. This indicates that allowing for placing the charging stations at the middle points reduces the number of charging stations compared to the case when they are restricted to being placed at the nodal points, and hence, contradicts the optimality of having the stations at the nodal points. As a final case, consider a vehicle with an AER less than 75 miles, in which case no possible nodal combinations will refuel the trip, and having stations at middle points are necessary [10].

The FRLM takes a set of O-D pairs and the associated traffic volumes as inputs. For setting charging station locations under FRLM, one must enumerate all different combinations of facilities that can charge the vehicle for a round trip. Each combination is evaluated by the following mixed integer programming to maximize the traffic captured by the stations [10].

$$\begin{aligned}
\text{maximize} \quad & \sum_{q\in Q} f_q c_q \\
\text{s.t.} \quad & \sum_{h\in H} u_{qh} v_h \geq c_q, \ \forall q \in Q \\
& e_{hn} a_n \geq v_h, \ \forall h \in H; e|e_{hn} = 1 \\
& \sum_{n\in N} a_n = P \\
& a_n, v_h, c_q \in \{0,1\}, \ \forall n \in N, \ \forall h \in H, \ q \in Q
\end{aligned} \tag{10.7}$$

The definitions of the optimization parameters are as follows:

Q : set of all O-D pairs in a given region

q : index of O-D pairs

f_q : PEV traffic volume along O-D path q

$$c_q : \begin{cases} 1, & \text{if flow } f_q \text{ is captured} \\ 0, & \text{otherwise} \end{cases}$$

N : set of potential charging station locations

n : index of potential facility locations

$$a_n : \begin{cases} 1, & \text{if a station is located at n} \\ 0, & \text{otherwise} \end{cases}$$

P : total number of stations to be sited

H : set of all potential station combinations

h : index of set H

$$e_{hn} : \begin{cases} 1, & \text{if station n is in combination of h} \\ 0, & \text{otherwise} \end{cases}$$

$$u_{qh} : \begin{cases} 1, & \text{if station combination h can charge O-D pair q} \\ 0, & \text{otherwise} \end{cases}$$

$$v_h : \begin{cases} 1, & \text{if all facilities in combination h are open} \\ 0, & \text{otherwise} \end{cases}$$

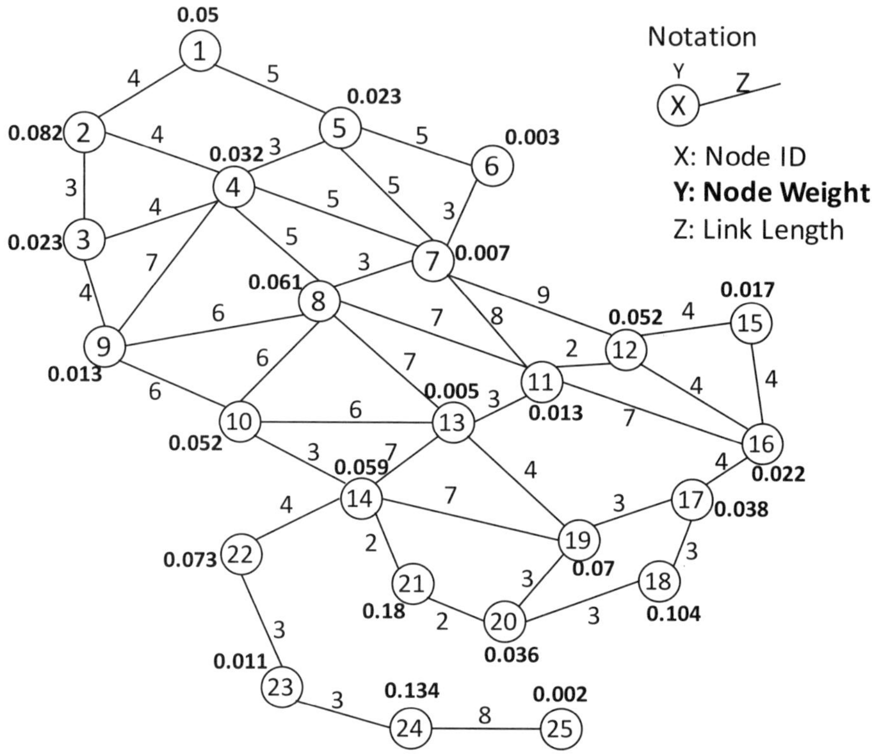

Figure 10.9 A 25-node test network, also used in [9, 10, 18, 19].

The objective function in (10.7) aims to serve as many PEVs as possible with P stations, as enforced by the third constraint. The first constraint states that at least one eligible facility combination h should be able to serve the path. Similarly, the second constraint holds v_h to 0, unless all the stations in combination h are serving customers and the last constraint is the integrality requirement.

10.4.1 Case Study

The model mentioned above is tested on a 25-node network given in Figure 10.9. In this network, the distances between the nodes are written on the edges, e.g.,

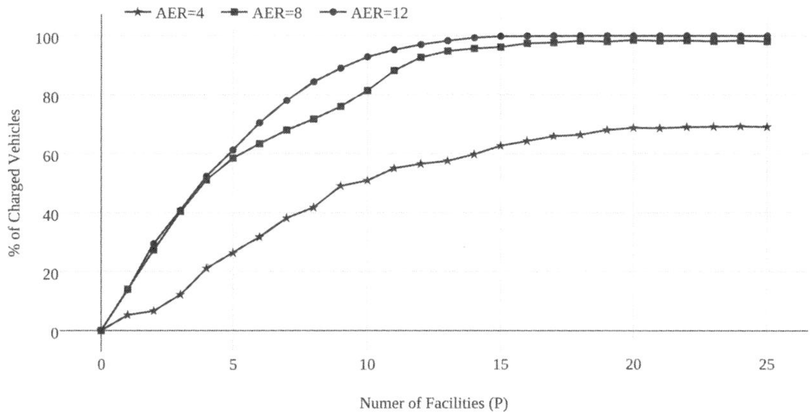

Figure 10.10 Trade-off between % of charged vehicles and number of stations for different AER [10].

the distance between nodes 1 and 5 is five units, and the node weights are written in bold. The traffic volumes between the nodes i and j are estimated through the gravity model:

$$OD_{ij} = \frac{W_i W_j}{d_{ij}^2} \tag{10.8}$$

where W_i is the weight of the origin node i, W_j is the weight of the destination node j, and d_{ij} is the shortest path between nodes i and j. In the next step, the shortest paths between the nodes are calculated. As reported in [10], the driving distance affects the size of the problem. For instance, for a driving range of 4 units, there are 300 continuous flow coverage variables (c_q), 75 continuous fractional variables (v_h), and 25 integer variables (e_n). When the driving range increases, there is more combination of charging stations, hence the problem grows.

It is important to note that vehicle range has a significant impact on facility location decisions. For instance, for a vehicle range of 4, locating stations to all of the 25 networks will not charge all of the vehicle flow, as the distance between several nodes (e.g., nodes 10-13, nodes 13-14, etc.) is farther than 4 units. The main reason is that there are many direct links with a distance longer than 4 (e.g., 1 to 5, and 5 to 6). To serve all the traffic demand, the maximum node distance should be less than or equal to the driving distance. The charging station optimization problem is solved for three different AER: 4, 8, and 12 driving units. As shown in Figure 10.10, as vehicle range increases, the number of stations to cover the same amount of traffic flow decreases.

To illustrate the exact locations of the stations, suppose that the AER is 4 units. For the case of a single station ($P = 1$), the station will be located at node 21. The main reason is that this node can charge PEVs coming from neighboring nodes as the direct distance 2 units and the traffic flow is higher than the other candidate sites (e.g., 20, 18, and 23). As a second case, suppose $P = 2$, in which case, the stations will be located at nodes 18 and 20. In parallel to the previous argument, for the case of $P = 3$ stations, the optimal locations are nodes 18, 20, and 21.

The example provided above has three shortcomings. First, the model assumes that the charging stations can serve the entire demand irrespective of how large the traffic flow is. This might be valid at the early stages of PEV integration, when a few chargers can handle the low demand traffic. In fact, recall from previous chapters that the station has zero blocking probability for certain low arrival rates. Therefore, the models may need to include *station capacity*. Such problems are called *capaciated* FRLM [20]. Second, the framework only targets to maximize traffic that can be captured, and it ignores the underlying power system dynamics as some locations may not be suitable for deploying stations due to lack of infrastructure. Finally, the model does not capture the stochasticities involved with PEV movements. Therefore, the station models may need to include the facility location provided in the previous chapters.

10.5 Chapter Summary

In this chapter, we presented and discussed the challenges associated with designing the locations of the charging stations, which has a key role in PEV integration. In the first part of the chapter, we presented an overview of facility location theory, which has been applied to many engineering and optimization problems. Next, we showed the distinguishing features of the slow-charging and fast-charging facility location problems. The public slow-charging option naturally requires vehicles to be stationary for several hours. Hence, slow-chargers will be deployed at currently existing parking lots. By considering the fact that both demand and candidate charging sites are at the discrete location, this problem was formulated via the P-median framework (P-center or set covering could also be applied). The demand for chargers was linked to the zone-level parking demand, which was calculated based on a travel survey conducted in Seattle, WA. In traditional fuel capturing or fuel refueling location problem, the goal is to place facilities in a given network, with weighted nodes representing traffic intensities and edges representing the distances. The objective is to maximize the captured traffic. Most of the time it is assumed that stations have infinite capacity and a PEV is assumed to be charged when it passes by a station. On the other hand, deploying large-scale fast-charging stations is likely to be prohibited by the power grid constraints. Therefore, optimization frameworks must consider the cases in which customer demands are served by limited station resources. Under such circumstances, facility location theory should additionally involve station parameters such as waiting times and blocking or loss probabilities.

In the future, the placement of charging stations will possibly involve two decision variables. The first is the location of stations and the second is the number of chargers to be deployed at the stations. The objective is to minimize blocking probabilities (as defined in the previous chapters), or to maximize the captured traffic, so that PEVs driving between the origin-destination pairs can complete the trip. In this sense, stations located at the high traffic volumes may need to limit the state of charging levels for vehicles, but allow vehicles to complete their trips. As discussed in the previous chapter, this approach reduces service durations and stations can charge more batteries. Therefore, determining the SoC levels for each station could be a possible third decision variable.

References

[1] Chen, T. D., Kockelman, K. M., and Khan, M., "The Electric Vehicle Charging Station Location Problem: A Parking-Based Assignment Method For Seattle," *Proc. Transportation Research Board 92nd Annual Meeting*, Washington DC, Jan. 2014, pp. 13-1254.

[2] ReVelle, C. S. and Eiselt, H. A., "Location Analysis: A Synthesis and Survey," *European Journal of Operational Research*, Vol. 165, No. 1, 2005, pp. 1-19.

[3] Owen, S. H., and Daskin, M. S., "Strategic Facility Location: A review", *European Journal of Operational Research*, Vol. 111, No. 3, 1998, pp. 423-447.

[4] Melo, M. T., Nickel, S., & Saldanha-da-Gama, F., "Facility Location and Supply Chain Management A Review," *European Journal Of Operational Research*, Vol. 196, No. 2, 2009, pp. 401-412.

[5] Daskin, M. S., "What You Should Know About Location Modeling," *Naval Research Logistics (NRL)*, Vol. 55, No. 4, 2008, pp. 283-294.

[6] Daskin, M. S., *Network and Discrete Location: Models, Algorithms, and Applications*, John Wiley & Sons, 2011.

[7] Arabani, A. B., and Farahani, R. Z., "Facility Location Dynamics: An Overview of Classifications and Applications," *Computers & Industrial Engineering*, Vol. 62, No. 1, pp. 408-420, 2012.

[8] Wesolowsky, G. O., "Dynamic Facility Location," *Management Science*, Vol. 19, No. 11, 1973, pp.1241-1248.

[9] Hodgson, M. J., "A Flow Capturing Location Allocation Model," *Geographical Analysis*, Vol. 22, No. 3, 1990, pp. 270-279.

[10] Kuby, M., and Lim, S., "The Flow-Refueling Location Problem for Alternative-Fuel Vehicles," *Socio-Economic Planning Sciences*, Vol. 39, No. 2, 2005, pp. 125-145.

[11] Fetene, G. M., et al., "The Economics of Workplace Charging," *Transportation Research Part B: Methodological*, Vol. 88, 2016, pp. 93-118.

[12] Frade, I., et al., (2011). "Optimal Location of Charging Stations For Electric Vehicles in A Neighborhood In Lisbon, Portugal," *Transportation Research Record: Journal of the Transportation Research Board*, No. 2252, pp. 91-98.

[13] Liu, Z., Wen, F., and Ledwich, G., "Optimal Planning of Electric-Vehicle Charging Stations in Distribution Systems," *IEEE Transactions on Power Delivery*, Vol. 28, No. 1, 2013, pp. 102-110.

[14] Lam, A. Y., Leung, Y. W., and Chu, X., "Electric Vehicle Charging Station Placement: Formulation, Complexity, And Solutions," *IEEE Transactions on Smart Grid*, Vol. 5, No. 6, 2014, pp. 2846-2856.

[15] Pugent Sound Regional Council Maps http://www.psrc.org/data/gis/shapefiles.

[16] Pugent Sound Regional Council Maps http://www.psrc.org/data/transportation/travel-surveys.

[17] Bayram, I. S., et al., "On the Evaluation of Plug-in Electric Vehicle Data of a Campus Charging Network," *IEEE International Energy Conference*, Leuven, Belgium, Apr. 2016, pp. 1-6.

[18] Cruz-Zambrano, M., et al.,"Optimal Location Of Fast Charging Stations in Barcelona: A Flow-Capturing Approach," *Proc. 10th International Conference on the European Energy Market*, Stockholm, Sweden, May 2013, pp. 1-6.

[19] Simchi-Levi, D., and Berman, O., "A Heuristic Algorithm For The Traveling Salesman Location Problem On Networks," *Operations Research*, Vol. 36, No. 3, 1988, pp. 478-484.

[20] Upchurch, C., Kuby, M., and Lim, S., "A Model for Location of Capacitated Alternative Fuel Stations," *Geographical Analysis*, Vol. 41, No. 1, 2009, pp. 85-106.

Chapter 11

Load Balancing in Charging Station Networks

11.1 Introduction

In order to ensure satisfying consumer driving-range needs, and to address the seemingly paradoxical connection between driving-range anxiety and technology adoption trends, as discussed in the previous chapter, a network of DC fast-charging infrastructure is required. In fact, statistics from late 2016 show that on California highways nearly 300 fast chargers were already deployed along with several thousand slow chargers at public locations such as restaurants and hotels. Moreover, the U.S. Department of Energy is investing $4.5 billion to build a charging network ("coast-to-coast, nationwide zero-emission network") on the highways to boost the penetration rate.

While deploying new chargers is necessary to match the growing demand, growing the size of the charging infrastructures will face limitations imposed by power grid constraints, such as line capacities, congestion, and transformer loadings. One solution to circumvent the operational limits is to make the network of charging stations distributed in order to meet PEV demand. However, as discussed previously, station locations and customer demand may not be perfectly aligned, hence there is a need to develop a strategy to route customers. As discussed in the studies in [1] and [2], such routing can be facilitated through a financial incentive mechanism. In routing schemes, customer preferences are modeled by a utility function that represents the relation between the incentives and the willingness to drive extra miles. To that end, this chapter presents a Stackelberg game-based PEV routing framework, in which the network operator acts as the leader and sets the prices at each station, while PEV drivers, who act as the followers, choose the charging facilities so that they minimize their costs. The key features of the proposed framework include the following items:

1. The stations are operated by one entity, hence the goal is to coordinate the demand. This assumption is based on the fact that stations located in geographically well-defined regions are typically served by a single utility company,[1] hence locational marginal prices are similar across stations.

2. Due to short charging sessions, demand cannot be split and must be served by one station.

3. Each station is hard-capacitated (i.e., has a finite serving capacity).

Stackelberg game-based modeling of charging networks has received some limited attention [3] and [4]. The study in [3] proposes a competition among stations, wherein each station sets prices to influence customers. The study in [3] provides a closed-form solution for a two-station model and larger networks are not treated due to the complexity of the problem.

11.2 Network of Fast-Charging Stations

In this section, we develop a fast-charging station framework by building on the model presented in Chapter 7, which proposed a fast-charging station equipped with local energy storage units. Chapter 7 also introduced a stochastic model for customer arrival rates and service durations in order to quantify service performance based on service blocking rates or loss-of-load probabilities. As discussed in the previous chapter, a network of fast-charging stations is of particular importance to facilitate longer commutes, such as interstate trips in the United States or cross-country travel in Europe. In this section, we present the key component of the network charging station and a decentralized load balancing framework, which is referred to as the PEV admission control (PEVAC) mechanism.

1 This is mostly the case in the United States; however, European energy markets may have multiple utilities serving a region.

11.2.1 PEV Admission Control

As stated in previous chapters, congestion pricing is an effective tool for coordinating and managing customer demands. So far in this book, we have applied various pricing schemes to single-charging stations. In this section, however, we use pricing mechanisms to manage the PEV load across the entire network. It is assumed that PEV drivers are self-interested, rational, and act independently, meaning that they are interested in receiving service with minimum cost without considering charging network conditions. Therefore, stations located at busy routes may experience high customer demands, while neighboring charging stations may remain underutilized. This situation not only leads to economic losses and inefficient customer service, but also may result in system failures. Such situations can be avoided by employing admission control mechanisms, which aim to influence the station selection choice by offering financial incentives. Overall, the goals of the PEVAC can be listed as follows:

- Attain network-wide predefined quality of service targets.
- Act as a tool to balance the PEV demands among neighboring stations located within a close proximity.
- Maximize the network's revenues by serving more customers and efficiently using grid resources.

PEVAC uses a pricing mechanism that dynamically adjusts the prices of charging services according to actual load on the network. For instance, during rush hours, station operators can offer relatively lower prices in underutilized stations to shave excessive demand at a busier one.

11.2.1.1 System Parameters

Next, we review the dynamics of the pricing-based admission control mechanism employed at the station level. During the mild demand conditions (e.g., zero blocking or no congestion) it is assumed that each PEV chooses to go to the nearest charging station. In the PEVAC model (shown in Figure 11.1), the arrival rates of the PEVs are represented by $\lambda_{\text{PEV}}(t)$. Each arriving customer enters the pricing block, and $\lambda_{\text{ac}}(t)$ is the arrival rate that accepts the offered price. Similarly, $\lambda_{\text{r}}(t)$ is the arrival rate for routed customers from the neighboring stations and $\lambda_{\text{ad}}(t)$ is the arrival rate of the admitted PEVs. On the other hand, some PEVs may visit busy stations, in which case they might be denied service. To distinguish customer

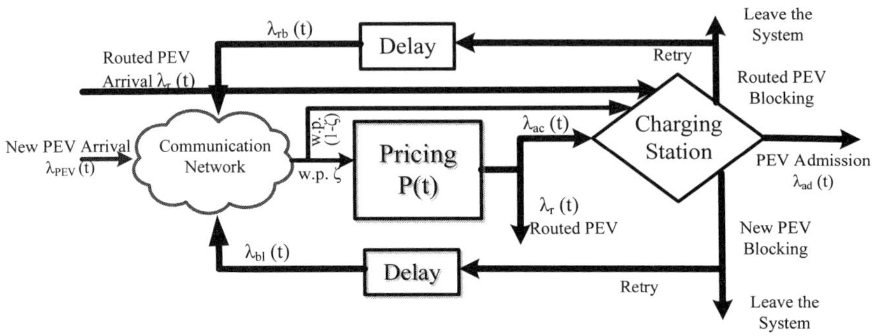

Figure 11.1 Plug-in electric vehicle admission control (PEVAC).

weights, two types of PEV blocking probability are defined: (1) B_{PEV} as the blocking of PEVs who come to the nearest station, and (2) B_{rb} as the blocking of a routed PEV. Furthermore, when a blocking occurs, it is assumed that the blocked customer will wait for a random duration and try the pricing block again. Overall, it is expected that one-third of customers retry to receive service and the rest leave. $\lambda_{\mathrm{bl}}(t)$ and $\lambda_{\mathrm{rb}}(t)$ represent arrival rates of customers who retry the system. The list of system parameters are given in Table 11.1.

11.2.1.2 Pricing Block and QoS

The blocking probability, also referred as the loss-of-load probability in the previous chapters, represents the percentage of customers denied service in the long run, and represents the QoS metric. The performance metric is denoted by P_{BT}, as the weighted sum of the two blocking types; that is, $P_{\mathrm{BT}} = \gamma_1 B_{\mathrm{PEV}} + \gamma_2 B_{\mathrm{rb}}$, where $\gamma_1 + \gamma_2 = 1$. Since blocking a PEV that is routed from a neighboring station results in higher customer dissatisfaction, it is assumed that $\gamma_2 > \gamma_1$.

The pricing block is the critical component of the PEVAC mechanism. Recall that pricing policies in blocking systems are grouped into three categories [5]. The schemes in the first category use a *flat pricing* scheme, according to which customers are charged by their service duration. The main advantage of such policies is their straightforward implementation, which makes the accounting processes simple. On the other hand, flat pricing schemes do not consider the state of the system and fail to alleviate congestion. The second group of pricing schemes employs *dynamic pricing*, which monitors the system load continuously to avoid congestion by adjusting the prices. However, in the case of charging stations, implementing such pricing policies is rather practically challenging, since they require deploying advanced real-time monitoring and measurement infrastructures.

The framework presented in this section employs a *myopic policy*, which falls in between the flat and the dynamic pricing schemes. Specifically, the pricing block $P(t)$ works as follows. Let

$$\tilde{\lambda}_{\rm PEV} = \lambda_{\rm PEV}(t) + \lambda_{\rm bl}(t) + \lambda_{\rm rb}(t) \ . \tag{11.1}$$

Unless the customer arrival rate $\tilde{\lambda}_{\rm PEV}(t)$ exceeds a certain threshold, say $\lambda^*_{\rm PEV}$, which is set by the QoS target, the station offers normal price $p_{\rm normal}$, which is the same for all stations. It is further assumed that $p_{\rm normal}$ is acceptable by each customer. During a congestion period, when QoS target is violated $\tilde{\lambda}_{\rm PEV}(t) > \lambda^*_{\rm PEV}$, the system operator adjusts prices accordingly and offers *congestion price* $p_c > p_{\rm normal}$, so that PEVs will potentially prefer to go to the neighboring (also less busy) stations. Notice that the busy-hour price can vary based on the actual customer load on the system.

Charging stations employed in the charging network are described in Chapter 7. Recall that the customers are served by a grid with capacity C, energy storage size R^2, and storage technology parameter ν. The optimal arrival rate is defined as the maximum arrival level at which the charging facility performance (blocking probabilities) stays within the QoS target $\delta_{\max}$. This can be calculated based on

$$\lambda^*_{\rm PEV} = \begin{cases} \max & \tilde{\lambda}_{\rm PEV} \\ \text{s.t.} & \mathbb{P}_{\rm BT}(\tilde{\lambda}_{\rm PEV}) \le \delta_{\max} \end{cases} . \tag{11.2}$$

It is important to note that the QoS target $\delta_{\max}$ is specified in the service level agreement (SLA), hence it is an exogenous variable. Next, let us define the pricing policy employed for computing the pricing block $P(t)$. As depicted in Figure 11.1, the pricing block $P(t)$ determines the percentage of customers who accept the offered price at time t. That is

$$[\lambda_{\rm PEV}(t) + \lambda_{\rm bl}(t) + \lambda_{\rm rb}(t)]P(t) = \lambda_{\rm ac} \ , \tag{11.3}$$

2 We used B to represent battery size in Chapter 7. It is modified in this chapter to avoid confusion with the blocking probabilities.

which yields

$$P(t) = \frac{\lambda_{\mathrm{ac}}}{\lambda_{\mathrm{PEV}} + \lambda_{\mathrm{bl}} + \lambda_{\mathrm{rb}}} \leq \frac{\lambda^*_{\mathrm{PEV}}}{\tilde{\lambda}_{\mathrm{PEV}}} \,. \tag{11.4}$$

The equations above show that the fraction of PEVs that accept the offered price is inversely proportional to the aggregate load on the station. Hence, as the arrival rate $\tilde{\lambda}_{\mathrm{PEV}}$ increases, the station operator increases the prices, and consequently, fewer customers will accept the price. Note that the demand function satisfies $D(p_{\mathrm{normal}}) = 1$, signifying the fact that all customers agree on this price. It is also well known that the demand functions describe the sensitivity of customers to the price changes. Hence,

$$p(t) = D^{-1}\left(\frac{\lambda^*_{\mathrm{PEV}}}{\lambda_{\mathrm{PEV}} + \lambda_{\mathrm{bl}} + \lambda_{\mathrm{rb}}}\right) , \tag{11.5}$$

where

$$\frac{\lambda^*_{\mathrm{PEV}}}{\lambda_{\mathrm{PEV}} + \lambda_{\mathrm{bl}} + \lambda_{\mathrm{rb}}}$$

is the load on the system. Various demand functions — for services providing QoS in congested networks — have been proposed in the literature. In this chapter, we use the demand function proposed in [1]. Therefore, $p(t)$, which is the price at time t, becomes

$$p(t) = \begin{cases} p_{\mathrm{normal}} \,, & \text{if } \tilde{\lambda}_{\mathrm{PEV}}(t) \leq \lambda^*_{\mathrm{PEV}} \\ p_c = p_{\mathrm{normal}}\left\{1 + \theta\sqrt{-\log\left(\frac{\lambda^*_{\mathrm{PEV}}}{\lambda_{\mathrm{PEV}} + \lambda_{\mathrm{bl}} + \lambda_{\mathrm{rb}}}\right)}\right\} , & \\ & \text{if } \tilde{\lambda}_{\mathrm{PEV}}(t) > \lambda^*_{\mathrm{PEV}} \end{cases} \,. \tag{11.6}$$

When congestion occurs in a charging station (i.e., $\tilde{\lambda}_{\mathrm{PEV}}(t) > \lambda^*_{\mathrm{PEV}}$), resource pricing according to customer load will improve the QoS. Parameter θ is set by the utility operator, the details of which will be explained in the next section. The interdependence among system components is presented in Figure 11.2.

11.2.2 Resource Allocation Framework

In order to serve more customers with the limited station capacities, system operators also need to consider spatiotemporal demand variations and allocate more system resources to stations located at busier regions. Let us assume that N charging

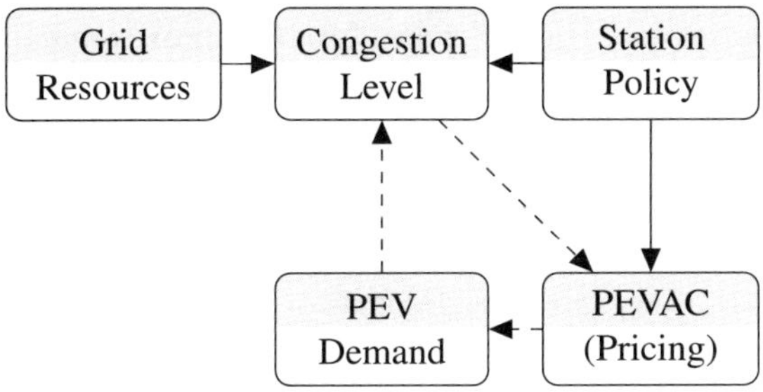

Figure 11.2 Interaction of system diagrams.

facilities are deployed in a serving area, and $\mathcal{N} = \{1, \ldots, , N\}$ represents the set of the charging stations. Furthermore, let $C_{\max}$ denote the maximum portion of generation capacity that the utility can provide to the charging network. It is assumed that each station is equipped with storage units. Hence, the problem simplifies to power resource allocation under QoS constraints. Specifically, the network operator solves the following two-phase resource allocation problem.

Phase I

In the first phase, for any given set of $\{R_i\}_{i \in \mathcal{N}}$ and $\{\lambda_i\}_{i \in \mathcal{N}}$ we solve the following problem:

$$\min_{C} \quad \sum_{i \in \mathcal{N}} P_{\mathrm{BT}}^{(i)}(\lambda_i, C_i, R_i) \tag{11.7}$$

$$\text{s.t.} \quad \sum_{i \in \mathcal{N}} C_i \leq C_{\max} \tag{11.8}$$

$$\delta_{\min} \leq P_{\mathrm{BT}}^{(i)}(C_i, R_i, \lambda_i) \leq \delta_{\max} \tag{11.9}$$

$$C_i \in \mathbb{Z}+, \tag{11.10}$$

$$\delta_{\min}, \ \delta_{\max} > 0 \tag{11.11}$$

The arrival rate λ_i for each charging station $i \in \mathcal{N}$ is calculated via discrete event simulation [6] for a given spatial distribution of vehicles (discussed in the next section). Also note that a lower limit is enforced on the the blocking probability targets as given in (11.9). The goal of setting a lower limit, which is typically set to

$\delta_{\min} = 0.01$, is to prevent system planners from overprovisioning the capacity. The solution of the above optimization problem will give the system operator a power allocation vector $\boldsymbol{C} = [C_1, \ldots, C_N]$, but in reality the implementation of this result may be constrained by the power grid operational limits. Suppose that a station located in a heavy traffic junction requires $C = 15$ units. Considering the fact that a typical DC fast charger draws 50 kW of power, the deployment of such a station may be improbable. Therefore, an additional constraint on the aggregate capacity is enforced. Based on this, the excess power

$$C^{\text{excess}} = C_{\max} - \sum_{i \in \mathcal{N}} C_i \,, \tag{11.12}$$

is distributed among the neighboring charging stations inversely proportional to the physical distance between stations. By following this distribution strategy, the demand will be satisfied by the closest charging facility via the pricing-based routing framework introduced in the previous section.

Phase II

Next, let us define $\Psi \subset \mathcal{N}$ as the subset charging of facilities that have excess power resulted from *Phase I*. Suppose that station $j \in \Psi$ has excess power C_j^{excess}. Then in Phase-II a neighboring station $k \in \mathcal{N} \setminus \Psi$ receives additional C_{kj}^{extra} amount of power from charging station j. Hence, the second phase of the allocation problem can be formulated as

$$C_{kj}^{\text{extra}} = \lceil C_j^{\text{excess}} \omega_k \rceil \,, \quad \text{where } \omega_k = \frac{1}{d_{kj}^2} \left(\sum_{z \in \mathcal{N} \setminus \Psi} \frac{1}{d_{zk}^2} \right)^{-1} \,, \tag{11.13}$$

where d_{zj} is the physical distance between stations z and j. A simple example is discussed next to further clarify (11.13). Assume that four stations serve a neighborhood, and Station 1 has six units of excess power. The remaining three stations do not have any excess power and they will share six units of power according to (11.13). Let us assume that the distance between Station 1 and Stations 2, 3, and 4 are $d_{21} = 1$, $d_{31} = 2$, and $d_{41} = 2$. Then, the excess power is allocated as follows: $C_{21}^{\text{extra}} = 4$, $C_{31}^{\text{extra}} = 1$, and $C_{41}^{\text{extra}} = 1$. Since Station 2 is the closest one to Station 1, during a congestion period customers can be routed to Station 2 with the least amount of necessary incentives.

Table 11.1
Systems Parameters

Parameter	Definition
$\lambda_{\mathrm{PEV}}(t)$	PEV arrival rate to the nearest station.
$\lambda_{\mathrm{ac}}(t)$	PEV arrival rate who accepted the price at the nearest station.
$\lambda_{\mathrm{r}}(t)$	Routed PEV arrival rate of routed (from neighboring stations).
$\lambda_{\mathrm{bl}}(t)$	PEV arrival rate of blocked customers.
$\lambda_{\mathrm{rb}}(t)$	PEV arrival rate of blocked routed customers.
B_{PEV}	Blocking probability of customers (who pick the nearest station).
B_{rb}	Blocking probability of the routed customers.
p_{normal}	Normal price accepted by every driver. The same price at each station.
$p_{\mathrm{congestion}}$	Congestion price depends on the station load.
p_{B}	Penalty paid for blocking a customer.
δ	QoS targets, which may be different at each station.
λ^{*}	Maximum allowable arrival rate to support QoS calculated by (11.2).
$c_{k_{\mathrm{inctv}}}$	Minimum amount of desired savings to accept routing.
$c_{k_{\mathrm{near}}}$	Cost to drive to nearest station.
d_k	Distance vector for each customer k.
θ	Vector of price tuning parameter for each station set by station operator.
C, R, μ	Single-charging station parameters. See Chapter 7 for details.
C^{limit}	Upper power limit that can be drawn by a single station.

11.3 Decentralized Control of Charging Stations

In this section, we present the decentralized control framework, which aims to serve a large population of PEVs under QoS guarantees without requiring significant upgrades. To that end, we propose a game-theoretic PEV framework to balance PEV demand among the nearby charging stations. The players of the game include the network operator and the PEV drivers. The primary objective of the self-interested PEV drivers is to charge their vehicles within QoS guarantees and to minimize the cost of service, which is proportional to the charging price and the driving distance to the charging facility. On the other hand, the system operator aims to maximize the number of served customers by offering incentives to PEV drivers to cooperate and route customer demand to nearby stations with relatively lower utilization, if necessary.

As depicted in Figure 11.1, the proposed framework relies on the capability of both parties to communicate. It is assumed that each charging facility can communicate with a central unit that can forward pricing information to mobile PEVs, and influence their preferences. On the other hand, the customers respond to these signals and act according to the best response strategy.

11.3.1 Game Formulation

In the proposed framework, the charging network owner acts as the *leader* who can commit to a strategy before the *followers* (PEVs) can select their strategies. To that end, a Stackelberg game (also known as a leader-follower game [1, 2]) is used to model the system. First, the game is defined in its strategic form $\Gamma = \{\mathcal{N}, \mathcal{K}, \{\boldsymbol{p}_k\}, \{X_k\}_{k \in \mathcal{K}}, \{U_n\}_{n \in \mathcal{N}}, \{U_k\}_{k \in \mathcal{K}}$, where $\mathcal{K} = \{1, \cdots, K\}$ the set of PEVs that require charging at a given time. The strategies of each set of players are given as follows. The set $\{\boldsymbol{p}_k\}$ denotes a $1 \times N$ price vector offered by the system operator to customer k. Note that in order to set the pricing vector, the network operator needs to set $\boldsymbol{\theta}$ for each node. The rationale behind setting parameter $\boldsymbol{\theta}$ stems from the fact that convincing drivers to change their charging station preferences requires that the ratio of peak-to-off-peak hour (or congestion hour to noncongestion hour) should be larger than one [7]. Therefore, the parameter $\boldsymbol{\theta}$ is set to maintain such incentives for customers. The PEV driver's strategy, X_k, is to pick a charging station from $\mathcal{N}$. U_k is the utility function that represents the payoff for player k. The game in its extensive form is also shown in Figure 11.3. Next, we describe the components of the game in detail.

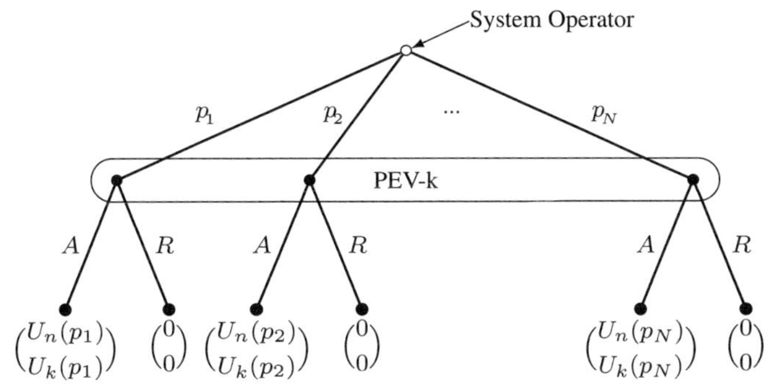

Figure 11.3 Stackelberg game in its extensive form.

11.3.2 PEV Drivers (Followers)

As mentioned earlier, during off-peak hours each PEV will simply choose to go to the nearest charging station. However, when the load at each station increases, the network operator aims to route the PEVs to other nearby stations with lighter loads (if necessary) by offering relatively lower prices, while still avoiding penalties for poor QoS (blocked PEVs). The drivers can either *accept* to go to a less busy charging station or *reject* and go to the nearest one and take the risk of getting blocked. Obviously, the decisions of the drivers are influenced by the price responsiveness. Hence, enough incentives should be presented to motivate the PEVs to change their charging behavior [8]. Considering this, the PEV strategy is given below.

$$\text{PEV Strategy} = \begin{cases} \text{Accept,} \quad \text{if} \begin{cases} c_{k_{\text{near}}} - c_{k_{\text{desrd}}} \geq c_{k_{\text{inctv}}} \\ P_{\text{BT}}^{\text{desired}} \leq \delta_{\max} \end{cases} \\ \\ \text{Reject,} \qquad\qquad \text{otherwise.} \end{cases} \tag{11.14}$$

The components of the PEV strategy are as follows. Suppose that customer k, for $k \in \mathcal{K}$ plans to charge his or her battery at the nearest station n, for $n \in \mathcal{N}$. The cost of choosing the nearest station is $c_{k_{\text{near}}} = p_n + c\left(d_{nk}^2\right)$, where p_n is the price paid for the service at station n, and $c\left(d_{nk}^2\right)$ is the cost of driving to station n for PEV k. Note that the charging price p_n equals to p_{normal} when there is no congestion (i.e., $\tilde{\lambda}_{\text{PEV}}(t) \leq \lambda^*_{\text{PEV}}$) at the nearest station n, and to the congestion price p_c during congestion as given in (11.6). Similarly, when the network operator needs to route a PEV to another station, it costs $c_{k_{\text{desrd}}}$. As mentioned above, the incentives should be large enough for drivers to drive extra miles. To incorporate the customer behavior into our formulation, we assume that PEV k will only agree to go to the less-busy station if it gains at least $c_{k_{\text{inctv}}}$. For instance, considering current electricity prices there should be a reasonable level of savings (e.g., 10%-15% savings).

The pay-off of a single PEV is represented by a utility function that has the following components: the QoS metric $\boldsymbol{P}_{\text{BT}}$; the price at each station $\boldsymbol{p}_k$; and the physical distance to each station $\boldsymbol{d}_k$. Note that each component of the utility function corresponds to a $1 \times N$ row vector, where the position of each element is associated with the corresponding charging station parameter (QoS, service price, and distance). Then, the utility function for PEV k becomes:

$$U_k(\boldsymbol{P}_{\text{BT}}, \boldsymbol{d}_k, \boldsymbol{p}_k) = h(\boldsymbol{P}_{\text{BT}})\boldsymbol{e}_n \left\{\boldsymbol{p}_k + c_k(\boldsymbol{d}_k) + f_k(\boldsymbol{d}_k)\right\} , \tag{11.15}$$

where $h(\boldsymbol{P}_{\text{BT}}) = e^{\xi(\boldsymbol{P}_{BT}-\delta)}$. The function $h(\cdot)$ represents the disutility experienced due to high blocking probability [9], δ is the QoS target, and ξ is a constant that reflects how much the customer values the service (e.g., higher value for when there is an urgent need for charging services). Furthermore, parameter ξ is introduced to also provide some flexibility in accommodating PEVs who once had been denied service, but then retried requesting service. In such cases, without parameter ζ, and when the blocking probability is high, it can further penalize the disutility. These components capture the dissatisfaction introduced by a high blocking probability. Note that

$$\xi = \begin{cases} 0, & \lambda \leq \lambda^* \\ \xi \in \mathbb{R}^+, & \text{otherwise} \end{cases} . \tag{11.16}$$

PEV k picks one station $n \in \mathcal{N}$, and $\boldsymbol{e}_n$ represents a column vector comprising of all zeros except for the nth position, which is 1. Similarly, $\boldsymbol{p}_k$ is the pricing vector provided by the network operator. The second term in the utility function reflects the cost of driving to a charging station. Finally, $f_k(\boldsymbol{d}_k)$ represents the dissatisfaction of PEV k when the driver chooses to drive extra miles. Even though the total cost is lowered, this involves some level of dissatisfaction due to spending extra time to drive the extra miles (e.g., time cost). Therefore, the optimization problem of a PEV is a mapping $\mathbb{R} \to \{\text{reject}, \text{accept}\}$ that gives rise to a vector $\boldsymbol{e}_n$, where "accept" sets the nth position to 1, while all other remain 0 corresponding to "reject." Note that the objective is to minimize the overall cost. This can be expressed as follows.

$$\begin{aligned} \min_{n} \quad & h(\boldsymbol{P}_{\mathrm{BT}})\boldsymbol{e}_n \left\{\boldsymbol{p}_k + c_k(\boldsymbol{d}_k) + f_k(\boldsymbol{d}_k)\right\} \\ \text{s.t.} \quad & n \in \mathcal{N} \end{aligned} \tag{11.17}$$

The cost functions used in (11.17) are as follows. The cost of driving is $c_k(\boldsymbol{e}_n \boldsymbol{d}_k) = p_{dr} d_{kn}^2$, for which the quadratic cost functions are proposed in the classic analysis of Hotelling [10]. The other component captures the dissatisfaction of PEVs to drive extra miles. A linear dissatisfaction model is used: $f_k(\boldsymbol{e}_n \boldsymbol{d}) = p_{\mathrm{dis}}(d_{kn} - d_{\mathrm{nearest}})$, where p_{dis} is the cost of driving one unit of distance, and $(d_{kn} - d_{\mathrm{nearest}})$ is the total extra miles driven. This cost component aims to capture the behavior of drivers, because it is unlikely for a customer to drive a significant number of extra miles for limited savings. Finally, the above optimization problem is subject to PEVs having enough stored charge to drive to station n, which we assume is the case (about 1-2 kWh of stored energy).

11.3.3 Network Operator (Leader)

In the game setting, the network operator has the first move advantage to set the charging price at each station to influence the PEV preferences. Therefore, the network operator acts as the *leader* who aims to coordinate the behavior of the noncooperative customers so that it maximizes the profit of the charging network. The strategy of the leader is to tune the pricing parameter vector $\boldsymbol{\theta}$ at each station. Hence, the offered price vector becomes a function of vector $\boldsymbol{\theta}$ (i.e., $\boldsymbol{p}_k = h(\boldsymbol{\theta})$). Moreover, it is assumed that privacy policies forbid the system operator to take advantage of customer locations to ask for higher prices of those whose charge levels are low. Similarly, a network operator can discard customers who repeatedly try to reenter a congested station. Hence, the utility function of the network operator can be cast as

$$\begin{aligned} \max_{\boldsymbol{\theta}} \quad & \boldsymbol{p}_k(\theta)\boldsymbol{q} - p_B \boldsymbol{q}_B \\ \text{s.t.} \quad & \boldsymbol{p}_k, \boldsymbol{q},\ \boldsymbol{q}_B \in \mathbb{R}^{\mathrm{N}},\ p_B \in \mathbb{R}^{+}, \end{aligned} \tag{11.18}$$

where, $\boldsymbol{q}$ is calculated from (11.17), and equals to $\boldsymbol{e}_n$. Similarly,

$$q_B = \begin{cases} \mathbf{0}, & \text{if service received} \\ \boldsymbol{e}_n, & \text{if service denied.} \end{cases}, \tag{11.19}$$

and the price vector $\boldsymbol{p}_k$ is the same as the one in (11.17). Finally, p_B is a scalar and represents the price paid when blocking occurs.

11.4 Numerical Results

11.4.1 Spatial Vehicle Distribution

Recall that the main motivation for the aforementioned vehicle routing framework is the fact that spatial vehicle distribution is nonuniform, and as a consequence, the stations' serving capacity and customer demand may not align perfectly. On the other hand, collecting vehicular traffic traces is a challenging and costly task. In order to reflect traffic traces in our simulations, we use the traces of the bus movements in Seattle, WA. Due to the city's extensive bus network, 1200 buses in 5000 km square area, the available data is assumed to resemble actual vehicle patterns [6, 11]. Furthermore, we use the weekday rush hour data (periods 7 am-9 am and 5 pm-7 pm). ARENA Input Analyzer Software is used to fit the data to a spatial distribution. First, the location coordinates (x, y) are normalized and the results given below are calculated with a mean squared error of 0.6. The resulting piecewise distribution is given as

$$f(x) = \begin{cases} 44 \times Be(4.42, 0.763) & 0 \leq x \leq 44 \\ 44 + 137 \times Be(0.752, 4.7) & 44 \leq x \leq 180 \end{cases}, \tag{11.20}$$

and

$$f(y) = \begin{cases} 150 \times Be(2.42, 0.799) & 0 \leq y \leq 150 \\ 150 + 121 \times B_e(1.07, 5.44) & 150 \leq y \leq 270 \end{cases}, \tag{11.21}$$

where B_e denotes the Beta distribution. Moreover, the correlation coefficient of x and y coordinates is calculated as 0.06 [6].

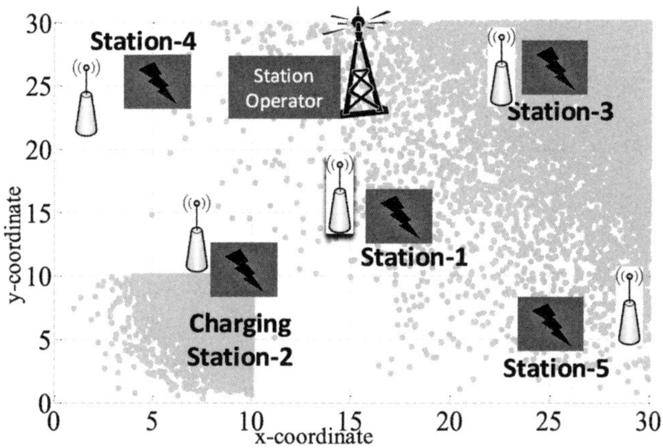

Figure 11.4 Charging network layout.

11.4.2 Simulation Setting

The application of the proposed customer routing framework is illustrated with the following simulation scenario. The charging network comprises five charging nodes in a 30×30 unit square area. The facility coordinates are $(5, 25)$, $(10, 10)$, $(25, 25)$, $(15, 15)$, and $(25, 5)$ for Stations 1 to 5, respectively. The station layout and the vehicular distribution are depicted in Figure 11.4. The locations are chosen to show the effectiveness of the routing framework, as Stations 2 and 3 are under heavy traffic, while other stations serve light traffic regions. The physical distance between Station 1 and Stations 2 and 3 is shorter than the distances between Stations 2 and 3 (busy) and Stations 4 and 5 (less occupied). More precisely, the total demand in terms of percentages is 1%, 50%, 42%, 2%, and 5% for Stations $1-5$, respectively. In this case, Stations 2 and 3 will experience high customer rejection rates, while the other stations perform relatively well.

The station settings are as follows. The utility company provides $C_{\max} = 39$ and each station can receive $C^{\text{limit}} = 13$. According to (11.7) and (11.13), the optimal resource allocation yields $\boldsymbol{C} = [6,\ 13,\ 13,\ 3,\ 4]$ for the five stations in the network. Notice that the excess power at Stations 2 and 3, i.e., $C_2^{\text{excess}} = 5$ and $C_3^{\text{excess}} = 2$, is transferred to Stations 1, 4, and 5 as [5, 3, 1]. The charge rates are $\mu_1 = \cdots = \mu_5 = 2$, while the local energy storage unit parameters are identical and set to $\nu_i = 3$. Also, the size of each storage unit is $R_i = 8$, for all $i \in \mathcal{N}$.

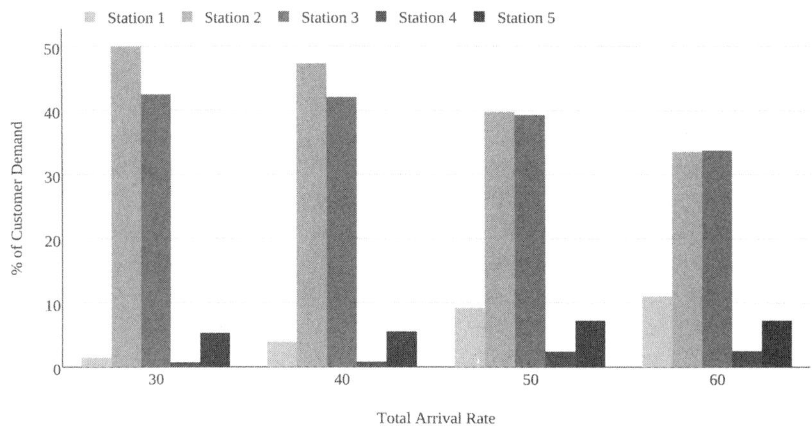

Figure 11.5 Load balancing in a network of charging stations.

The blocking target is set as $\delta = 0.05$, and customer dissatisfaction parameter is set as $\xi = 0.1$ for all customers. Given the stations' settings, the optimal arrival rates given in (11.2) are calculated as $\boldsymbol{\lambda}_i^* = [6.7,\ 23.4,\ 23.4,\ 3.3,\ 5.0]$. When the arrival rates exceed the optimal rates, the station starts to offer incentives in order to route customers. The service price during off-peak periods is $p_{\text{normal}} = 4$, whereas if a PEV is blocked, the station operator pays a penalty of $p_{\text{block}} = 6$. The coefficients for two types of blockings are set as $\gamma_1 = 0.45$ and $\gamma_2 = 0.55$. Drivers may have different incentive levels, hence $c_{k_{\text{inctv}}}$ is a random variable with uniform distribution in the interval $[0.75, 1.0]$, and p_{dis} is a uniformly distributed random number (per unit of distance) in the interval $[0.02, 0.05]$.

To illustrate the customer routing framework, the network is simulated for aggregate arrival rates in the range [30,60]. Under light traffic regime, Stations 2 and 3 can provide charging services within the specified QoS limits. However, as demand increases, more customers are routed to Stations 1, 4, and 5. This trend is depicted in Figure 11.5, where it is shown that 25% of customer demand is routed from busy stations to nearby stations. Moreover, the blocking probability performance is evaluated for three cases; namely, a baseline scenario in which the stations are identical and no power allocation or customer routing policy is implemented, a power allocation framework, and a power allocation and customer

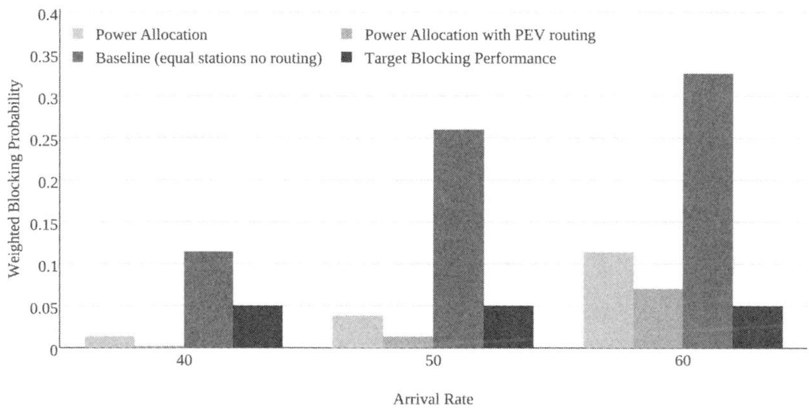

Figure 11.6 Charging network performance: Weighted blocking probability comparison for three cases.

routing framework. As shown in Figure 11.6, the proposed model outperforms the other two cases and serves more customers. It is also important to note that under high traffic regimes (e.g., $\sum_i \lambda_i = 60$), the QoS policy is slightly violated because there is a need to upgrade energy storage capacity or power drawn from the grid. The last evaluation is on comparing the system revenue (explained in Section 7.3) for the aforementioned three cases. The results presented in Figure 11.7 show that the network operator makes more profit by serving more vehicles, despite the offered incentives.

11.5 Chapter Summary

In this chapter, we introduced a pricing-based customer routing framework to match the demand and response pools for large populations of mobile PEVs. The proposed decentralized mechanism relies on the Stackelberg game in which the station operator acts as the *leader* and sets the prices at each station to influence customer decisions in order to serve more demand. On the other hand, customers act as the followers and aim to reduce their cost by selecting their serving stations. The

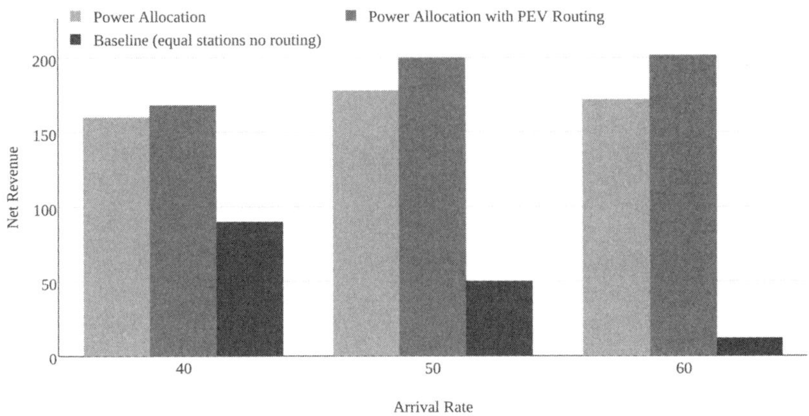

Figure 11.7 Network-wide revenue comparison.

numerical experiments showed that more PEV demand can be satisfied by routing customers to less busy stations. Routing frameworks are critical elements in grid planning since they can reduce the need for additional power generation.

References

[1] Bayram, I. S., Michailidis, G., and Devetsikiotis, M., "Unsplittable Load Balancing in a Network Of Charging Stations Under Qos Guarantees," *IEEE Transactions on Smart Grid*, Vol. 6, No. 3, 2015, pp. 1292-1302.

[2] Bayram, I. S., et al., "Decentralized Control Of Electric Vehicles in a Network of Fast Charging Stations," *IEEE Global Communications Conference*, Atlanta, GA, Dec. 2013, pp. 2785-2790.

[3] Yuan, W., Huang, J., and Zhang, Y. J. A., "Competitive Charging Station Pricing for Plug-in Electric Vehicles," *Proc. IEEE International Conference on Smart Grid Communications*, Venice, Italy, Nov. 2014, pp. 668-673.

[4] Yang, H., Xie, X., and Vasilakos, A. V., "Noncooperative and Cooperative Optimization of Electric Vehicle Charging Under Demand Uncertainty: A Robust Stackelberg Game," *IEEE Transactions on Vehicular Technology*, Vol. 65, No. 3, 2016, pp. 1043-1058.

[5] Hampshire, R. C., Massey, W. A., and Wang, Q., "Dynamic Pricing to Control Loss Systems with Quality of Service Targets," *Probability in the Engineering and Informational Sciences*, Vol. 23, No. 2, 2009, pp. 357-383.

[6] Bayram, I.S., et al., "Electric Power Allocation in a Network of Fast Charging Stations," *IEEE Journal on Selected Areas in Communications*, Vol. 31, No. 7, 2013, pp.1235-1246.

[7] Faruqui, A., & Sergici, S., "Household Response to Dynamic Pricing of Electricity: A Survey Of 15 Experiments," *Journal of Regulatory Economics*, Vol. 38, No. 2, 2010, pp. 193-225.

[8] Faruqui, A., et al., "Will Smart Prices Induce Smart Charging of Electric Vehicles?", 2011, available at SSRN 1915658.

[9] Lau, V. K., and Maric, S. V., "Mobility of Queued Call Requests of a New Call-Queueing Technique For Cellular Systems," *IEEE Transactions on Vehicular Technology*, Vol. 47, No. 2, 1998, pp. 480-488.

[10] Hotelling, H., "Stability in Competition," *The Collected Economics Articles of Harold Hotelling*, 1990, pp. 50-63.

[11] Jetcheva, J. G., et al, "Design and Evaluation of a Metropolitan Area Multitier Wireless Ad Hoc Network Architecture," *Proc. Fifth IEEE Workshop on Mobile Computing Systems and Applications*, Monterey, CA, Oct. 2003, pp. 32-43.

Chapter 12

Vehicle-to-X Systems

12.1 Introduction

In order to reduce the cost of PEV ownership, the idea of taking advantage of PEV batteries as a power resource during peak hours, blackouts, and other emergency situations has been receiving growing interest. The technologies that rely on PEV batteries as power sources are referred to as vehicle-to-X (V2X) technologies, which signifies the direction of the electricity from the PEV batteries to X, where X can represent the grid (G), a building (B), a home (H), or other PEVs (V).

In this chapter, we present a detailed overview of V2G, V2B, V2H, and V2V applications, their system architectures, and the associated cost and benefit analysis. We classify these technologies as front-of-meter and behind-the-meter technologies according to the nature of their interaction with the distribution and transmission networks. The V2G technology is considered a front-of-meter technology, since PEVs inject power back to the grid using distribution and transmission network. In the V2G technology, a PEV interacts with a dispatcher or a system operator for energy transactions in the ancillary services market. In contrast, V2B, V2H, and V2V technologies are considered as behind-the-meter applications. In these technologies, PEVs are located at customer premises and discharged battery power is used only for customer needs. In this group of technologies, the vehicle battery communicates with a local energy management system.

In another classification, these technologies can be sorted according to the amounts of energy exchanged. Due to the minimum power constraints of ancillary service contracts, a higher amount of energy will be supplied in the V2G technology, while the V2B, V2V, and V2H technologies require less amounts of energy, which is shaped by the nature of the loads they serve. In principle, more vehicles participate in a typical V2G system compared with the behind-the-meter systems. For instance, a typical V2B system used at a commercial building requires about 5-6 vehicles, while a V2H system is employed in a single household and requires 1-2 cars. In the next few sections, we discuss the details of each of these technologies.

12.2 Vehicle-to-Grid (V2G)

V2G technology has the potential to provide system operators with more flexibility and enhanced stability through enabling bidirectional energy exchange between PEVs and the power grid. When large numbers of PEVs are connected to the grid and they are aggregated as a single power source through software solutions, they could act as a large battery in the grid. From the systems operator's standpoint, V2G technology and its benefits are multifold. The stored battery energy can be used for serving active power applications such as increasing the spinning reserve in generation, peak load shaving, load leveling, and regulation. Furthermore, V2G technology can improve power quality, facilitate starting synchronous induction motors, active filtering, and general reactive power regulations. From the customers' perspective, the benefits manifest in monetary incentives. Also, customers can use PEV batteries to back up energy during the intervals in which the electricity price is lower, and use the stored energy during peak hours or blackout periods. It is noteworthy that PEVs remain stationary and parked most of the times. Hence, V2G operations do not interrupt customer use of their PEVs.

The notion of V2G technology is gradually being implemented and realized beyond being a proof of concept. In the summer of 2016, a consortium consisting of multiple companies created a V2G hub with 10 PEV slots (10-kW V2G units) to enable drivers to participate in ancillary services in Denmark [12]. V2G technology has particularly received interest in Denmark because Denmark generates a significant fraction of its electricity from wind energy, the operation of which improves significantly by integrating batteries into the system. In a similar effort, in the United Kingdom, Nissan has partnered with local utilities to deploy V2G stations, and in the United States, a showcase has been designed and demonstrated at the University of Delaware, which operates in the PJM market, the details of which are provided in Section 12.3.4. Experiments show that by establishing a sufficient level

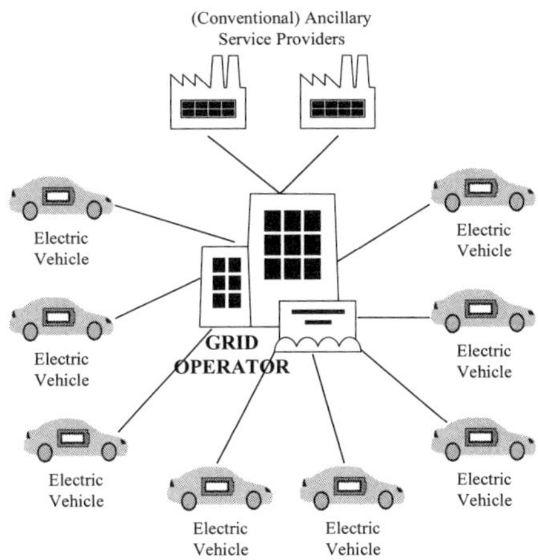

Figure 12.1 Illustration of a direct V2G system architecture: individual link is between the geographically dispersed PEVs and the grid operator [3].

of communication and information infrastructure, system operators can effectively control the process of power exchange between the grid and PEV batteries.

The V2G concept is realized by controlling and managing PEVs via data communication among the PEVs and an aggregator or a grid operator. The system architecture, which specifies the model of communication between the grid and the PEVs, can be either *direct* or *indirect* [3, 5, 6]. In direct architectures, as depicted in Figure 12.1, there are direct communication paths between the power grid operator and the individual PEVs. In this architecture, each participant can bid and perform ancillary services. Even though a direct architecture might seem simple enough to be implemented easily, there are challenges in implementing such architectures. Specifically, in conventional ancillary services, a system operator interacts with the contractors via automatic generation control (AGC) signals, which should be responded to by proper actions within one to several seconds. Furthermore, AGC is designed to communicate with large-scale power sources (e.g., 100 kW or higher) located in geographically well-confined regions. Therefore, the long-term scalability of direct communication architecture is an issue that should be addressed carefully, as the operator has to monitor and update the connection, contract, power, and the state of charge regularly for thousands of vehicles, which dynamically enter and leave the V2G system in real time.

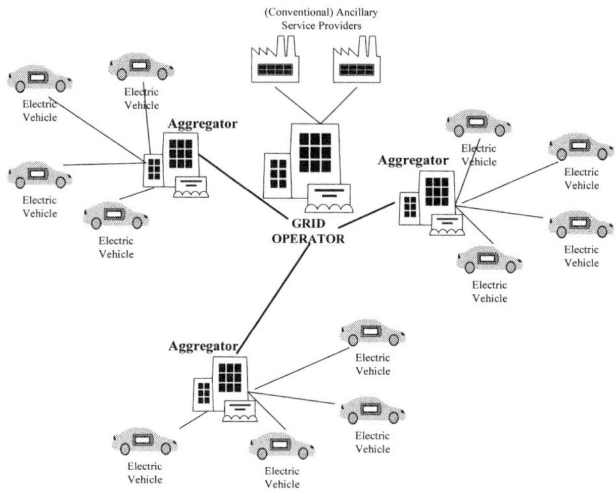

Figure 12.2 An illustration of an indirect V2G system architecture: PEVs are coordinated though aggregators. [3].

In contrast, in the indirect architectures, aggregator units are deployed, which are responsible for coordinating partitioned PEV groups to act as a single controllable power source. The system architecture is shown in Figure 12.2. In such architectures, aggregators bid and take part in the V2G system and issue commands to the contracted vehicles. Additionally, each aggregator manages a relatively fewer number of PEVs, which reduces the communication overhead significantly, reduces the complexity of scheduling and resource management, and facilitates a scalable architecture.

One of the critical requirements of any ancillary service provider is exhibiting a high degree of *availability*. Availability factor (AF) measures the fraction of the time that a generating unit provides services to power grids. In the traditional power generating units, the AF is affected by the fuel type as well as the design and age of the plant. These factors determine the frequency and duration of the outage events and necessary planned and unplanned maintenance procedures. According to the North American Electric Reliability Council, generation units should have an AF of 92.91% in order to meet ancillary services contract [3]. In V2G systems, availability is mainly affected by the probability that PEVs are available at the V2G stations, which in turn hinges on the number and duration of PEV trips during each day. The research conducted in [3] has studied the probability of a vehicle being parked at home or work based on data from the National Household Transportation Survey. This survey indicates that between 3 pm and 7 am, the probability of the vehicles being parked is at least 0.9, and moreover, the probability of being parked between 12 am to 4 am is 1. For other intervals, the probability varies between 0.83 and 0.90. Hence the average availability factor is at least 91%, which makes V2G systems a suitable candidate for ancillary services in the United States.

Reliability is another metric used to assess the suitability of V2G systems for participating in the ancillary service. In ancillary services, the system operator sends the AGC signals and asks the contractor (the PEV fleet) to supply the specified capacity for 10 minutes. Reliability is calculated as the fraction of the 10-minute periods during which the PEV fleet continuously supplies 100% of the binding power. Otherwise, the PEV fleet is considered to have failed to fulfill the contract. For instance, consider a scenario in which the PEV fleet fulfills the contract at every second for $n = 3$ 10-minute periods out of $N = 4$ contracted periods, in which case the reliability becomes $\frac{3}{4} = 75\%$. Similar measures are also used to quantify the performance of frequency regulation processes (explained in the following section) and the corresponding performance rewards and payments. According to California System Operator (CAISO) tariffs, "The CAISO will sum a resource's automatic generation control set points for each four second regulation interval every fifteen minutes, and then aggregate the total deviations from the automatic generation control set point for each four second regulation interval during that fifteen minute period. The CAISO will divide the sum of the resource's automatic generation control set points less the sum of the resource's total deviations by the sum of the resource's automatic generation control set points. The CAISO will apply the resulting percentage to a resource's instructed mileage to calculate the resource's regulation performance payments" [7]. Even though the rules and regulations of the ancillary services market are clear for conventional power generators, the rules may need to be changed in order to accommodate the V2G systems. For instance, the typical minimum amount of power for ancillary services is 1 MW. Providing such a minimum amount by the PEV fleet is not viable in the near future. Hence, both the minimum power amounts to be generated by PEVs and the corresponding rewards need to be reduced in order to promote integrating V2G systems.

12.3 Power System Ancillary Services: An Overview

Besides serving the electricity needs of end users, power system operators also provide a range of ancillary services to enhance the system reliability. The Federal Energy Regulatory Commission of the United States defines ancillary services as "necessary to support the transmission of electric power from seller to purchaser given the obligations of control areas and transmitting utilities within those control areas to maintain reliable operations of the interconnected transmission system." Ancillary services are indispensable parts of the power system operations and are responsible for balancing the supply with the demand in real time within tight tolerance bounds. The types of services and durations of the support may differ across different markets depending on operational regulations, properties of the load profiles, and generation assets. In current markets, which are more deregulated, ancillary services are typically run by independent entities such as regional transmission operators (RTOs) or independent system operators (ISOs).[1] Broadly, the ancillary services pertinent to PEV operations can be divided into two groups: frequency regulation and reserves (spinning and non-spinning).

12.3.1 Regulation

The most common ancillary service is frequency regulation (also known as active regulation). Frequency regulation is performed to keep the grid frequency within narrow tolerance bounds (e.g., within 0.1% of frequencies 50 or 60 Hz). Although it is still not feasible to store electricity in large volumes, supply must match the demand instantaneously and continuously with high precision. However, unanticipated fluctuations [2] at the supply and/or demand side will impact the system frequency. As depicted in Figure 12.3, when the generation is higher than the demand (overgeneration), system frequency increases. On the other hand, if the demand is higher than the supply, system frequency decreases. The fluctuations in system frequency lead to inter area oscillations and system blackouts (e.g., the 2003 blackout in the Northeast United States). Frequency regulation is maintained by the control actions described below. In the sequel, for describing the challenges associated with frequency regulation, we adopt the terminology used by the PJM market since PJM

1 For instance, there are seven such entities in the United States: California ISO, Midwest ISO, New England ISO, New York ISO, Ontario ISO, PJM (Pennsylvania, Jersey, Maryland) RTO, and Electric Reliability Council of Texas (ERCOT) [25].

2 Supply-side fluctuations arise due to equipment failures, while demand-side fluctuations occur due to varying human activities.

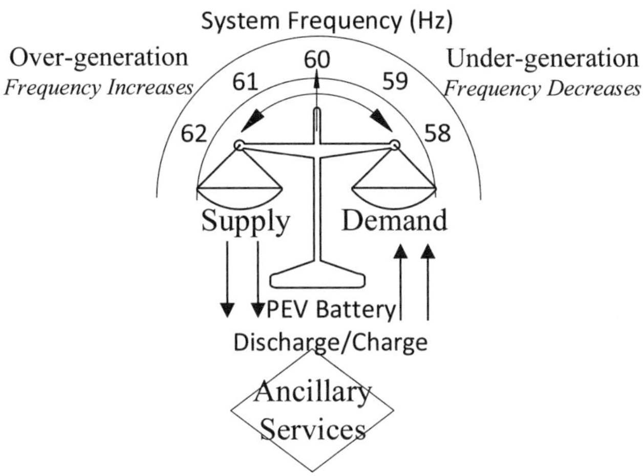

Figure 12.3 Frequency regulation and PEV-based ancillary services [25].

has the largest ancillary services market around the globe. This terminology might have differences with other regions or countries.

- Inertial response (IR): Large synchronous generators resist the instantaneous changes in the supply-demand equilibrium. IR occurs naturally, and it is uncontrollable. Generators slow down when the load increases and speed up when the load decreases.

- Frequency bias (FB): FB is the system operator's obligation to inject or absorb energy within a timeframe with a duration of 1 to 10 seconds. Frequency bias has two components, the governor response and the inductive load response. Governor response is the change of the generator output by a few MWs, and the inductive load response pertains to the changes in industrial inductive (motor) loads in response to the change in frequency.

- Regulation control (RC): PEVs could take part in regulation control in which the system operator purchases power from the ancillary services market. Suppliers bid for the impending time periods for the amount of power and the unit cost. The grid operator sorts the bids and determines the market-clearing prices at which the required supply is met at the lowest cost. Frequency

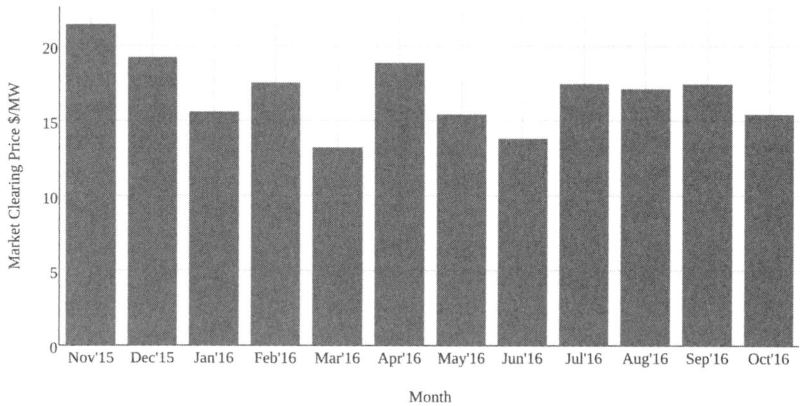

Figure 12.4 Average market clearing prices in PJM for frequency regulation.

regulation has two types. One type is called fast regulation, according to which regulation signals are generated within tens of seconds, and therefore, called fast regulation. In the second type, regulation occurs in minutes, hence called slow regulation.

- Economic control (EC): EC is used when the change in the load cannot be compensated for by the previous actions listed above. In such circumstances, the system operator moves the output of generators to a new set point. EC actions typically take over 5 minutes.

Participating in frequency regulation provides the PEV owners with financial incentives. The value and incentive for participating depend on the market clearing price. To provide more details, the average market clearing prices for the period of November 2015 through October 2016 are presented in Figure 12.4. The data presented in this figure is obtained from PJM database [13]. Even though the mean market price is $18/MW, the hourly rates fluctuate widely around that average in various seasons. The prices are generally higher during the peak hours and remain low during the off-peak hours. In 2011 the average market value for California, ERCOT, New York, MISO, and New England regions were $16.1/MW, $31.3/MW, $11.8/MW, $10.8/MW, and $7.2/MW, respectively.

12.3.2 Reserves

Reserves are supplementary generation capacity used to compensate for the loss of generation or load forecasting errors. Reserves are segmented into two categories: spinning (synchronized) or non-spinning (non-synchronized) reserves. Spinning reserves are actively generating units and have the ability to adjust their outputs according to market signals. Non-spinning reserves, on the other hand, are off-line generators, but they are fast-response generators that can provide power when needed. Generally, run-of-the-river hydro, pumped hydro, and combustion turbines are used for non-spinning reserves. Sometimes black start services are considered as a part of non-spinning reserves. Black-start generators provide the output when there is a blackout to restore the grid.

The response time for operating reserves is generally less than 10 minutes, and the duration of support is typically up to 2 hours. PEV batteries can provide such quick response time and low total energy. Furthermore, reserves require less charge and discharge requests compared with regulation (e.g., 25 – 50) times a year, which is slow enough to not degrade battery lifetimes [6]. The market value for spinning reserves is usually less than the frequency response. In 2011 the average market values for PJM, California, ERCOT, New York, MISO, and New England regions were \$10/MW, \$7.2/MW, \$22.9/MW, \$7.4/MW, \$2.8/MW, \$1.0/MW, respectively.

12.3.3 Motor Starting

Starting several large induction motors concurrently requires rapid injection of large amounts of reactive power [6]. If the reactive power is drawn from the distribution network, motor starting could lead to disruptions, such as voltage drops, in the nearby buses of the network. Voltage drops degrade voltage quality and possibly damage customers' appliances. A voltage drop of 3% leads to the flickering of lights and dropping more than 6% is not allowed by system operators. The extent of the voltage drop during a motor starting phase is a function of inrush current and the impedance of the power system and the motor. PEVs can be used to reduce the voltage sags caused by motor starting. PEV chargers with AC/DC inverters can provide reactive power to aid the motor starting process. Providing reactive power locally is advantageous to power system operations because the transmission of reactive power is eliminated, hence distribution lines are not congested. The use of AC/DC inverters also reduces the need to employ startup capacitors, which are used for hard starting of induction motors. Exclusion of capacitors will reduce the system

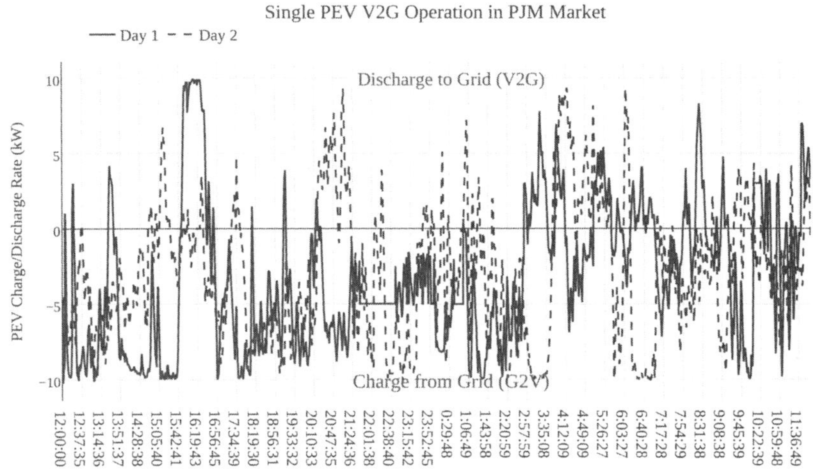

Figure 12.5 Charge/discharge rates of a V2G implementation at PJM ancillary market in 2013 at University of Delaware. (Data from [10].)

design cost and minimize the over-voltage risk that typically occurs if the capacitor is not disconnected after the motor starts off.

12.3.4 V2G Demonstrations at PJM and CAISO Regions

In April 2013, a partnership between the University of Delaware and PJM Interconnection allowed PEVs to serve the grid as an official power source. Through this partnership, software has been developed, which acts as a dispatcher to coordinate the charging and discharging rates of up to 15 PEVs with PJM. Due to minimum power requirements (100 kW) in the ancillary market, at least nine PEVs are used. The system operator uses vehicle batteries in a frequency response program to stabilize the grid. This method eliminates the need to cycle expensive thermal power plants, which are often more expensive, have slow response time, and have higher carbon emissions. On the other hand, each vehicle receives an incentive of \$150, while the electricity cost per car is about \$40. Hence, each participant PEV gains a profit of around \$110 per month[3].

3 The calculations exclude the cost of battery degradation, as it depends on the battery type, age, and other factors as explained in Chapter 4.

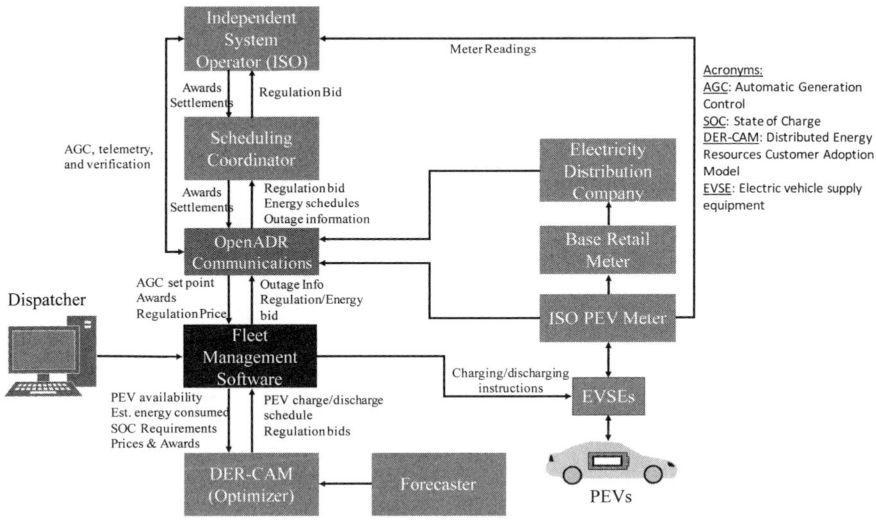

Figure 12.6 Vehicle-to-grid architecture of U.S. Department of Defense demonstration in California [3].

As discussed in the frequency Section 12.3.1, when electricity generation is more than the demand, PJM charges the PEV batteries for storage; when the demand is higher than the generation, the stored energy is drawn back to the grid. This, implicitly, relies on the assumption that the drivers have not drained their batteries completely. To illustrate the V2G operations, we provide the battery charge-discharge cycles of a single PEV for two days. In Figure 12.5, V2G operations are presented with a 30-second resolution. The upper side of the vertical axis represents the cases in which the PEV sells power to the grid, while the lower section shows the charge rate of the batteries. The graph shows that a PEV can take a proactive role in ancillary services and support stabilizing the grid.

The United States Department of Defense is also interested in developing V2G technologies for the energy security of military bases [11, 14]. In an experimental project at the Los Angeles Air Force base, 29 PEVs provide frequency regulation to CAISO. The station is composed of level 2 (up to 15 kW) and DC fast chargers (between 15 and 50 kW). The critical component of this model is the fleet scheduling, optimization, and control software developed by Berkeley National Laboratory to enable PEVs to participate in CAISO's ancillary services market. The

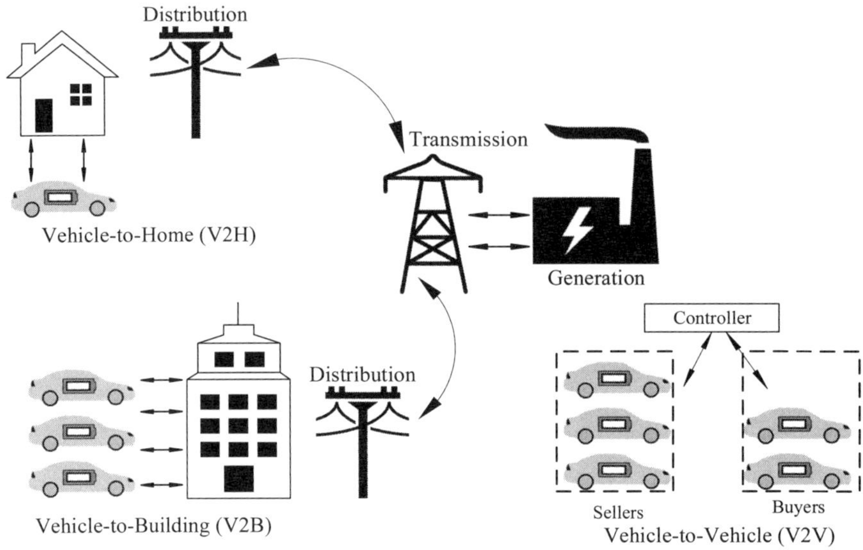

Figure 12.7 Overview of Behind-the-meter Applications.

overview of the software architecture is shown in Figure 12.6. The fleet management system[4] gathers the travel requirements and develops a schedule to jointly optimize PEV charging and regulation bid capacities. The bids and the rewards are communicated with CAISO using open-standard communication systems. Receiving and responding to dispatch instructions are enabled by the demand response automated server (DRAS). [5] DSAR receives CAISO signals and forwards them to the fleet management software via the Open Automated Demand Response Protocol (OpenADR). OpenADR is an open-source standard used to enable communications over IP-based networks between the electricity distribution units and system operators. OpenADR is also widely employed in demand-response programs. The optimization of fleet charge-discharge is carried on at the distributed energy resources customer adoption model (DER-CAM),[6] which is a decision-making and scheduling tool for renewable energy resources originally designed for microgrid planning. For

4 Bosch Software eMobility Solution fleet management software is used.

5 Developed by Akuacom Inc.

6 Developed by the Lawrence Berkeley National Laboratory.

managing the V2G systems, the key inputs to DER-CAM are PEV availability, estimated energy consumed, SoC requirement, electricity prices, and awards, while its outputs are PEV charge/discharge schedule and regulation bids.

12.4 Behind-the-Meter Applications

The primary focus of the V2G systems described in the previous section is to aid grid operations by charging and discharging the batteries. PEV batteries can also be deployed at behind-the-meter (at customer premises) and provide services such as reducing demand charge, integrating renewable energy sources, and reducing electricity costs. Such services can be further promoted when combined with additional policies that provide incentives to the customers for using them. For instance, the Self Generation Incentive Program in California offers \$1/W for advanced energy storage deployments at customer premises. There are three groups of systems that fall under this category of systems: namely vehicle-to-building, vehicle-to-home, and vehicle-to-vehicle systems. It is noteworthy that implementation of behind-the-meter systems is simpler than implementing V2G systems, since there is no interaction with the grid and the systems can be implemented in a relatively short duration. An overview of these systems is depicted in Figure 12.7.

12.4.1 Vehicle-to-Building

Vehicle-to-building (V2B) systems use multiple PEV batteries to satisfy the electricity demand of a building, especially during peak hours. V2B systems can also provide emergency backup and lower the energy cost of the buildings. In addition to energy charges, which are paid according to consumption, demand charges account for a sizable portion of customers cost. Demand charges are applied to most commercial and industrial customers, and to a lesser extent to residential customers[7]. Demand charges are applied to the customers who consume electricity at a higher rate (peak demand) since the utility company has to incorporate additional infrastructure (generation, transmission, and distribution) in order sustain its continuous service to such customers. For instance, in Northern California the consumption of commercial customers on average are over 15-minute intervals. If the maximum demand exceeds 499 kW, customers are charged with demand charges.

7 As of 2016, 19 utilities in the United States23941222 are using demand charges for residential customers.

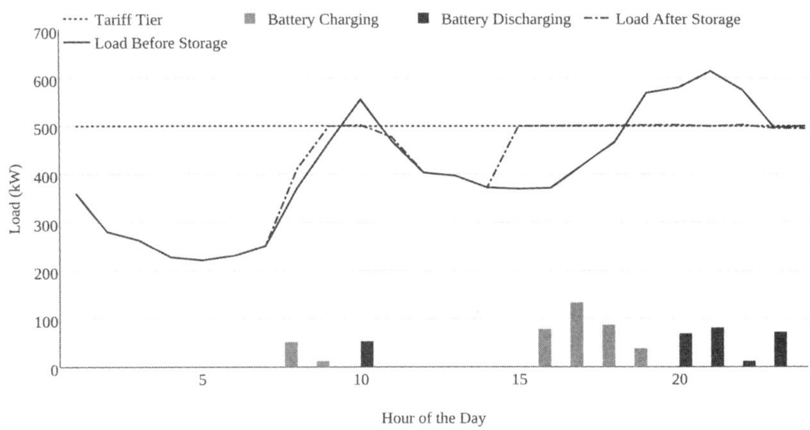

Figure 12.8 Vehicle-to-building application: Demand charge reduction in San Francisco, California [8].

V2B systems can be beneficial to commercial buildings, such as hotels and workplaces. In Figure 12.8, we show simulation results for a large hotel that uses several PEV batteries to charge and discharge during the day for avoiding demand charges. In this specific example, the system storage requirements are 140 kW and 560 kWh. PEV batteries are used two times during a typical day, once in the morning (9 am - 10 am) and another time during late evening (8 pm to 11 pm). For the case study, the hotel could lower the electricity cost by $20,000 annually [8]. It is noteworthy that the monetary savings depend on the pricing tariffs and the building load profile. Also, PEV batteries can engage in V2G activities to gain extra profit. Moreover, V2B systems can also support zero-energy buildings whose consumption is about the electricity generated by renewable sources on-site. Overall, the revenue gained by deploying V2B systems is due to the reduction in energy consumption and demand, and can be written as [17]

$$\text{Revenue} = \underbrace{E_{\text{shift}} \times (p_{\text{onpeak}} - p_{\text{offpeak}})}_{\text{Energy}} \times d + \underbrace{p_{\text{dcharge}} \times (P_{\max} - P_{\text{V2B}})}_{\text{Demand}}, \tag{12.1}$$

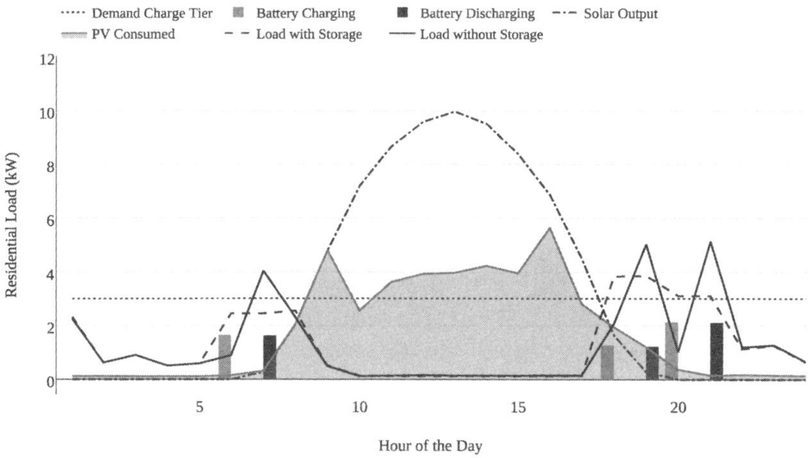

Figure 12.9 Vehicle-to-home application: Residential bill management and demand charges scenario in Phoenix, Arizona (Salt River Project tariff) [8].

where E_{shift} is the shifted energy (kWh) from on-peak to off-peak hours, p_{onpeak} and p_{offpeak} are the prices of electricity during on- and off-peak hours, d is the number of days in a month, p_{dcharge} is the unit cost of demand charges, $P_{\max}$ denotes the power demand before V2B serving the building, and P_{V2B} is the power demand after V2B. The output of V2B also depends on power electronics efficiency levels, and it can be stated as

$$P_{\text{V2B}} = P_{\text{vehicle}} \times \eta_{\text{charger}} \times \eta_{\text{inverter}} \times \eta_{\text{others}} \tag{12.2}$$

where P_{vehicle} is the output power of the PEV, while η_{charger}, η_{inverter}, and η_{others} are the efficiency levels of the bidirectional charger, DC to AC inverter, and other factors such as power loss and battery self-discharge, respectively. Note that (12.2) is also valid for V2G, V2H, and V2V systems. Moreover, in all V2X applications the battery degradation cost should be taken into account. The rate of battery degradation depends on factors such as the battery age, SoC level, DoD level, and ambient temperature. Hence, degradation cost changes case by case and the exact calculation depends on each case, and can often be complex. A detailed discussion of battery degradation is provided in Chapter 4.

12.4.2 Vehicle-to-Home

The second behind-the-meter system is the vehicle-to-home (V2H) system. Similar to the V2B system, PEV batteries are charged/discharged by bidirectional chargers. In V2H systems the outputs of the PEVs are not synchronized with the power grid, but alternatively, it provides power for off-grid AC loads. The difference between V2H and V2B is in their scales. A typical V2H system consists of one household and one PEV [15]. The primary application of the V2H systems is to use PEV batteries during grid outages or short-duration distribution system failures as an emergency backup to power the house. In this system, PEVs act as backup storage units and controllable loads. Adopting V2H systems will promote photovoltaic (PV) rooftop integration, reduce demand charges, and cut customers' electricity costs. V2H and V2B systems can further play important roles in the aftermath of the natural disasters (e.g., earthquakes and tornadoes) to provide emergency power supply [15].

A V2H system with PV generation, time-of-use pricing, and demand charges is discussed next. The quantifiable benefits of such systems depend on the household load profiles, solar intensity, the PV output, and most notably pricing tariffs. The simulation results shown in Figure 12.9 are assessed for a typical household and PV output profile in Arizona and the pricing policy is used from the Salt River Project (Arizona) tariff. In this pricing policy, in addition to the time-of-use pricing [8], a demand charge of $8.03 is applied if the demand exceeds 3 kW, $14.63 is applied if the demand exceeds the 7 kW threshold, and $27.77 is applied for additional kilowatts [6, 8]. In this case study, the goal is to increase local PV consumption and reduce the peak consumption inside the house. As depicted in Figure 12.9, the battery can be charged during off-peak hours and discharged during peak hours. It is noteworthy that the savings also rely on the PV output, demand profiles, and flexibility of appliances. For instance, the peak demand in a typical Japanese household is around 6 kW, while the peak demand in a two-bedroom apartment in the Gulf Cooperation Council (GCC) regions (e.g., Qatar and Kuwait) can be as high as 11 kW due to high demand for cooling. V2H architecture can also be used to reduce the cost of power outages and to improve power reliability. Table 12.1 summarizes the average cost of power outages in the United States. Power reliability is often measured by the following three metrics. System average interruption

8 (1) May, June, September, October TOU tariff: 1 pm - 8 pm (on-peak) 19.57 ¢/kWh, rest of the day (off-peak): 7.38 ¢/kWh. (2) July and August TOU tariff: 1 pm - 8 pm (on-peak) 22.26 ¢/kWh, rest of the day (off-peak): 7.41 ¢/kWh. (3) November-April TOU tariff: 5 am - 9 am and 5 pm - 9 pm (on-peak) 10.20 ¢/kWh, rest of the day (off-peak): 7.11 ¢/kWh.

Table 12.1
Average Outage Costs in the United States

$/kW	15 Minutes	30 Minutes	1 Hour	2 Hours	4 Hours	8 Hours
Residential	$0.05	$0.60	$2.60	$3.95	$5.30	$5.60
Small C&I‡	$8.65	$16.01	$23.37	$48.91	$117.76	$189.23
Large C&I	$4.79	$7.46	$10.12	$17.96	$36.94	$68.36

† Source: Electric Power Research Institute [9].
‡ Commercial and industrial customers.

frequency index (SAIFI) measures the frequency of outages, and is calculated by dividing the total number of outages per year by the total number of customers. The SAIFI index in the United States varies between 0.8 to 1.3 across the country with an average of 1.2. System average interruption duration index (SAIDI), on the other hand, measures the duration of the outages and is calculated as the total duration of the power outage divided by the number of customers. The SAIDI index in the United States ranges from 63 to 131, with an average of 106. In order to quantify the outages on customers, consumer average interruption duration index (CAIDI) captures the average number of minutes that a customer experiences outages in a year. CAIDI is calculated by dividing SAIFI by SAIDI. In the United States, the average CAIDI is 88 minutes per customer.

Both V2H and V2B systems require standards to enable hardware and software compatibility. Even though the industry lacks standards customized to these two systems, there are two related standards that can be applied to these systems. The first one is the IEEE 1547 standard, which defines the standards for interconnecting distribution resources with the electric power grid. Also, the SAE J3072 Standard provided by the Society of Automotive Engineers specifies the interconnection requirements and the communication needed for authorizing discharging of vehicles.

12.4.3 Vehicle-to-Vehicle

In vehicle-to-vehicle (V2V) systems, multiple PEVs exchange energy with each other when they are placed in a close vicinity, such as in parking lots. The process is managed through an aggregator or a controller. A typical application of V2V systems is when a PEV has an urgent need for charging during peak hours. In this circumstance, such a PEV can buy power from other PEV owners at an agreed rate. In V2V systems, there are multiple sellers and buyers, but the number of participants at each side is not necessarily the same. From an analytical standpoint, auction mechanisms could be used to model a framework in which the buyers post the amount of energy they need, and the sellers bid for the amount and unit cost of energy they are willing to trade [20]. Another scenario for using the V2V systems is during blackouts, where grid power is not available. In V2V power transfer, the state of charge levels and the expected time to depart from the location are the primary factors affecting the V2V operations. V2V operations also reduce the total amount of energy drawn from the grid, and increase the storage capacity of V2G systems by storing the extra energy generated from renewable resources, where the stored energy can be used in the V2V systems. The drawback of V2V systems is the energy conversion losses during the energy exchange among the vehicles. The interest in V2V applications has been sparse because of the low electricity cost, limited battery capacity, and the losses due to energy conversion. For instance, the study in [19] presents a distributed algorithm for V2V charging stations. The simulation results show that each PEV joining an energy exchange session as a seller makes a profit around \$0.1-\$0.15, which is relatively negligible. Hence, V2V applications are likely to occur only during emergency cases.

12.5 Chapter Summary

So far in the book, we have discussed optimal station designs, demand-side management tools for PEVs, and optimal station locations techniques. This chapter focused on reverse power transfer applications; namely, vehicle-to-grid, vehicle-to-building, vehicle-to-home, and vehicle-to-vehicle systems. Such applications are beneficial to both PEV owners and grid operators. We have shown that PEVs can participate in frequency regulation, spinning reserves, and other services like voltage support and motor starting. The references [3,5,6] have shown that most of the monetary benefits are due to participating in frequency regulation services. We have also presented case studies for the vehicle-to-building and vehicle-to-home systems. In both cases, the discharged battery energy is not sent back to the grid; rather it is consumed at customer premises. The primary goal has been reducing demand charges, promoting PV adoption, and reducing electricity costs. We have also provided some details about vehicle-to-vehicle systems; however, due to low electricity costs and energy losses during power conversion; such systems have limited applications.

References

[1] Tan, K. M., Ramachandaramurthy, V. K., and Yong, J. Y., "Integration of Electric Vehicles in Smart Grid: A Review on Vehicle to Grid Technologies and Optimization Techniques," *Renewable and Sustainable Energy Reviews*, Vol. 53, 2016, pp. 720-732.

[2] Ehsani, M., Falahi, M., and Lotfifard, S. (2012). "Vehicle to Grid Services: Potential and Applications," *Energies*, Vol. 5, No. 10, pp. 4076-4090.

[3] Quinn, C., Zimmerle, D., and Bradley, T. H., "The Effect of Communication Architecture on the Availability, Reliability, and Economics of Plug-In Hybrid Electric Vehicle-to-Grid Ancillary Services," *Journal of Power Sources*, Vol. 195, No. 5, 2010, pp. 1500-1509.

[4] Quinn, C., Zimmerle, D., and Bradley, T. H., "An Evaluation of State-of-Charge Limitations and Actuation Signal Energy Content on Plug-In Hybrid Electric Vehicle, Vehicle-to-Grid Reliability, and Economics," *IEEE Transactions on Smart Grid*, Vol. 3, No. 1, 2012, pp. 483-491.

[5] Zhou, Z., Levin, T., and Conzelmann, G., "Survey of US Ancillary Services Markets (No. ANL/ESD-16/1)," *Argonne National Laboratory (ANL)*, 2016.

[6] http://www.srpnet.com/prices/pdfx/April2015/E-27Pb.pdf.

[7] Conformed fifth replacement California ISO Tariff as of Oct 1 2016, https://www.caiso.com/rules/Pages/Regulatory/TariffArchive/Default.aspx.

[8] Fitzgerald, G., Mandel, J., Morris, J., and Touati, H., "The Economics of Battery Energy Storage," *Rocky Mountain Institute*, Boulder, CO, 2015.

[9] Rastler, D. M., "Electricity Energy Storage Technology Options: A White Paper Primer on Applications, Costs And Benefits," *Electric Power Research Institute*, 2010.

[10] University of Delaware, "Vehicle-To-Grid", http://www1.udel.edu/V2G/Tools.html.

[11] https://vehicle-grid.lbl.gov/project/los-angeles-air-force-base-vehicle-grid.

[12] Nissan, Enel and Nuvve Operate World's First Fully Commercial Vehicle-to-Grid Hub in Denmark http://www.enel.ru/en/events_and_news/news/15588/.

[13] http://www.pjm.com/markets-and-operations/ancillary-services.aspx.

[14] Marnay, C., "Los Angeles Air Force Base Vehicle to Grid Pilot Project," *Proc. ECEEE 2013 Summer Study on Energy Efficiency*, Hyeres, France, 3-8 June 2013.

[15] Tuttle, D. P., et al.,"Plug-In Vehicle to Home (V2H) Duration and Power Output Capability," *Proc. IEEE Transportation Electrification Conference and Expo*, Detroit, MI, Jun., 2013, pp. 1-7.

[16] Ancillotti, E., et al., "Using Electric Vehicles to Improve Building Energy Sustainability," *Proc. IEEE International Electric Vehicle Conference*, Florence, Italy, Dec., 2014, pp. 1-7.

[17] Pang, C., Dutta, P., and Kezunovic, M., "BEVs/PHEVs as Dispersed Energy Storage for V2B Uses in the Smart Grid," *IEEE Transactions on Smart Grid*, Vol. 3, No. 1, 2012, pp. 473-482.

[18] Ghaderi, A., and Nassiraei, A. A. F.,"The Economics of Using Electric Vehicles for Vehicle to Building Applications Considering the Effect of Battery Degradation," *Proc. Annual Conference of the IEEE Industrial Electronics Society*, Nov., 2015, pp. 003567-003572.

[19] Wang, M., et al.,"A Semi-Distributed V2V Fast Charging Strategy Based on Price Control," *Proc. IEEE Global Communications Conference*, Austin, TX, Dec., 2014, pp. 4550-4555.

[20] Wang, Y., et al., "A Game-Theoretic Approach to Energy Trading In The Smart Grid," *IEEE Transactions on Smart Grid*, Vol. 5, No. 3, 2014, 1439-1450.

Chapter 13

Integration Beyond Technology: Social and Economical Aspects

Despite ongoing incentive programs for promoting mainstream PEV acceptance, deployment of PEVs has not met the target rates set since 2010. Previous chapters focused on the technological challenges facing PEV adoption and the remedies for alleviating them. The purpose of this chapter is to review the social challenges facing PEV adoption. This is motivated by the fact that even though consumers are the key players in PEV adoption, technology designs often underestimate these roles. Specifically, social, economic, and political hurdles are as imperative as the technological ones to accelerate PEV adoption. For many consumers, PEV integration and the associated logistics and uncertainties lead to a significant shift in viewing transportation and its reliance on PEVs, which can significantly impact their decision regarding adopting a PEV. Social sciences can be instrumental in engaging end users and addressing such issues. For this purpose, in this chapter, we describe the major factors that determine the broader dissemination of PEVs, some theoretical tools explaining consumer behavior, and review policies that currently exist.

Chapter 3 discussed the technological barriers that hamper PEV adoption, such as high battery cost, limited cruising ranges, and the need for widespread charging infrastructures. These factors will have different levels of significance in different geographical or political regions. For instance, the southern regions of the United States are expected to suffer the most from the limited PEV driving range because of their vast areas, the layout of their cities, and their dependency on personal transportation because of relatively limited public transportation systems. On the other hand, in regions with carbon emissions reduction targets (e.g., California) the motivation for adopting PEV are stronger. Specifically, the study conducted in [1] uses the daily driving trips of motorists in the United States, and evaluates the current PEVs in the market based on their ability to meet the driving needs. The results indicate that 90% of travel needs can be satisfied by single overnight garage charging. In the United States, the transportation energy consumption per capita differs significantly across different states. Similarly, the work presented in [2] uses travel surveys with high-resolution GPS data to assess the capability of PEVs in supporting driver demands. The study in [2] evaluates millions of high-resolution trip profiles and concludes that even the more affordable PEVs can be used to support 87% of trips across different states. It is noteworthy that this rate is expected to be higher in European countries where the daily driving ranges are relatively shorter. From these studies, it is concluded that technological challenges should not be the main impediments toward mainstream PEV adoption. This conclusion, in conjunction with noting financial incentives offered by governments (discussed in Chapter 3), suggest that the PEV penetration rate should be significantly higher than the current rate. In this chapter, we seek to analyze other socioeconomic factors that affect the adoption rate.

Understanding customer socioeconomic concerns can guide policy makers to prioritize and adjust the incentive programs and the associated policies. This chapter is divided into three main sections. In Section 13.1, we provide a detailed overview of recent empirical studies conducted through field, online, and telephone surveys. The results depict the current perception of customers and their short-term purchasing behaviors. In Section 13.2, we provide an overview of the theoretical frameworks that have been used to analyze consumer behaviors and their intentions for adopting PEVs. The main emphasis will be placed on the *diffusion of innovations*, *theory of planned behavior*, and *rational choice theory*. In Section 13.3 we provide a case study on the previous two sections. An overview of the multidisciplinary nature of PEV adoption is illustrated in Figure 13.1.

13.1 Consumer Views on PEVs: Empirical Studies

To form appropriate policies that promote PEVs, it is essential to collect and analyze the opinions of the end users. The details of such opinions and perceptions are generally collected via online, door-to-door, or over-the-phone questionnaires and

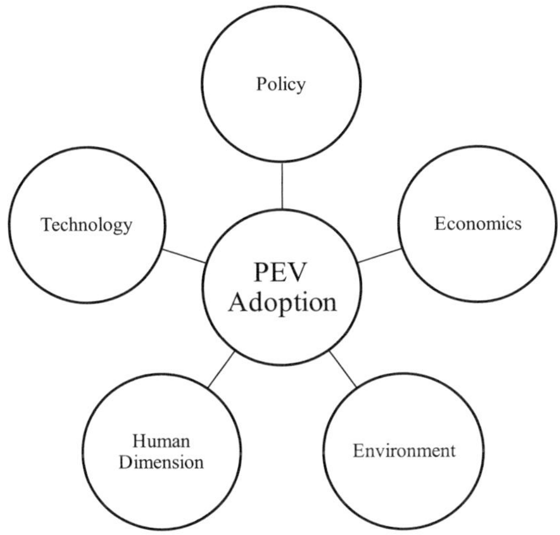

Figure 13.1 Multidisciplinary components of PEV adoption.

surveys. The scopes of such surveys and the ensuing analyses usually depend on the available resources for conducting the survey. Small-scale surveys are mostly conducted on campuses, at workplaces, or in communities, while large-scale surveys use sampling techniques to collect sufficiently diverse survey attendees that represent a society or a region.

One of the earliest survey-based studies was conducted in the United States in early fall 2011, which was before major manufacturers announced their PEV models [7]. The data was collected from 2302 individuals who were dispersed in 21 major urban areas. The survey primarily aimed to evaluate early awareness and intent of purchasing a PEV in the early stages. Driving range, charging duration, and the initial price were perceived as significant disadvantages by more than two-thirds of the respondents. On the other hand, the key advantages of PEVs included fuel economy and technological innovation, while environmental impacts of adopting PEVs were not among the major concerns. Survey attendants were asked to mark their likelihood of purchasing an electric vehicle on a scale of 1 to 10; where 1 and 10 represent the least and the most likely choices, respectively. The national average likelihood was 2.67, while the top regional average was 3.72 recorded in San Jose and San Francisco areas. Also, the studies revealed that early adopters are likely to be individuals who are more environmentally conscious.

Similar to this study, another extensive investigation on consumer perception about PEVs was carried out in the United States and reported in [4]. The study took place during February 2015 and involved 1015 households that are assumed to represent the United States population. Telephone interviews were conducted by Opinion Research Corporation commissioned by the National Renewable Energy Laboratories, and were focused on four main subjects that influence PEV penetration [4]:

1. *Vehicle purchasing behavior:* Vehicle ownership state of a household is a significant measure of the tendency toward acquiring a PEV. A household with multiple vehicles is likely to adopt PEVs because they can be used for daily trips and the conventional vehicles can be used for longer trips. The study revealed that 60% of households own more than two vehicles. Naturally, the type of vehicles that those in the household wish to drive is relevant as most PEV models are sedans. 53% of the attendants indicated that their last purchase was a sedan and 48% mentioned that their next purchase would be a sedan. The results showed that PEV manufacturers may need to introduce a diverse set of models to increase their market share. Furthermore, 59% of the respondents had bought a vehicle in the preceding three years and 50% of them were planning to purchase vehicles within the following three years. This trend anticipated a boost in PEV sales.

2. *PEV awareness:* To achieve high PEV sales, consumers should be informed of these technologies. The next part of the study examined PEV awareness and acceptance. Nearly 48% of the participators were able to name a PEV make and model (some top choices being Chevrolet Volt, Toyota Prius, Tesla S, and Nissan Leaf), while the remainder of the group could not. Similarly, respondents were asked whether they had seen a PEV. 57% of the attendants declared that they had been exposed to PEVs (have seen them in parking lots, had experienced sitting in a PEV, had driven one, or had neighbors with a PEV), while the rest of the population said they had never been near a PEV. These two questions are crucial as they describe how PEV adoption diffuses in the society and can be used in future studies that focus on analyzing the diffusion of innovations [8]. In order to assess customer acceptance, respondents were asked to compare PEVs with combustion engine models, and 52% of the respondents considered PEVs superior, while 45% considered both technologies to be similar.

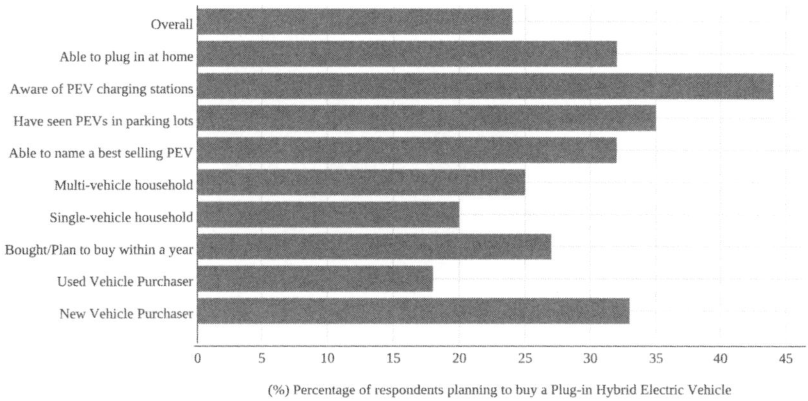

Figure 13.2 Willingness to consider buying a plug-in hybrid electric vehicle [4].

3. *Barriers to PEV acceptance:* In parallel to the technological barriers raised in Chapter 3, this survey also identified limited electric range, the availability of charging stations, and the initial battery cost as the major barriers for PEV adoption. To analyze customer perception about vehicle range, respondents were asked to specify a desirable driving range that would convince them to purchase a PEV. 27% of the people reported that they needed 1-100 miles, 14% of them required 101-200 miles, and 15% specified 201-300 miles. Thus to reach half of the population the vehicle battery should provide at least a 200-mile driving range.

   Regarding the availability of charging stations, which was identified as the second barrier, the questionnaire asked whether the respondents knew of any PEV charging stations along their daily trips. The vast majority of the respondents (79%) were unaware of any charging stations, while the rest of them had seen them at nearby locations such as stores and workplaces. Furthermore, 53% of the respondents stated that they had access to an existing electric outlet at their home. In the last question on this topic, the survey asked for the budget range that the respondents had for purchasing their next vehicle. The majority of the participants (70%) were willing to spend $30,000 or less, and 42% of respondents expected to pay $20,000 or

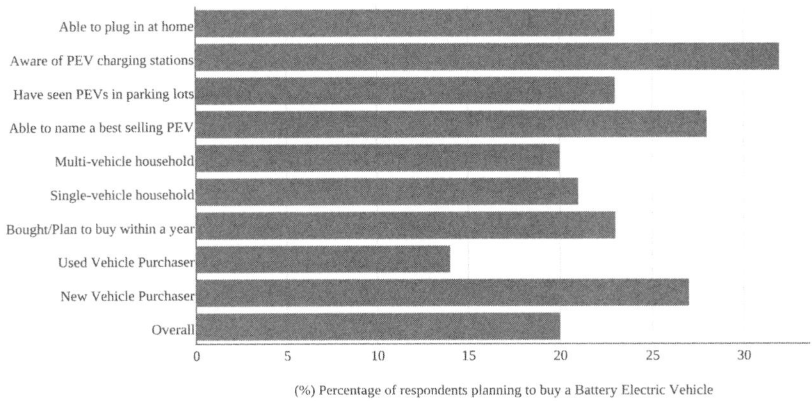

Figure 13.3 Willingness to consider buying a battery electric vehicle [4].

less. By incorporating the $7,500 tax credits offered by the United States federal government, Nissan Leaf would be in the budget range of 46% of the population, while the Chevrolet Volt would be in the purchasing range of 33% of the population.

4. *PEV acceptance:* The last part of the survey took the first three parts as inputs and aimed to understand the factors that promote PEV adoption. Respondents who were aware of public charging stations or had access to garage charging were more likely to purchase a PEV. Overall 24% of the population considered purchasing a PEV, while this number was 20% for battery electric vehicles. The characteristics of the respondents for each group are presented in Figures 13.2 and 13.3.

The studies and surveys conducted in the United States find that PEV awareness and acceptance rates have not improved, and have remained around 20%-27% during 2011-2015. The studies show that PEVs will not be attractive to mainstream consumers in the near future as PEV performance is compared against standard ICEs. Therefore, in [5] it is suggested that focusing on niche markets such as early adopters, postal cars, and car-sharing applications, is necessary for accelerating PEV adoption.

The next survey corroborates the same observations. In this survey, data was collected from a smaller group located at a technological university campus, comprised of students, faculty, and staff whom could be considered more likely to become adopters compared with the population average [10]. Awareness among the population is found to be higher than the general public since 53% of the respondents had some experience with alternative fuel vehicles. Moreover, 38% of the respondents indicated that they had high interest in PEVs, while 43% indicated that they had moderate interest in PEVs. The results show that policies focusing on early adaptors or technology enthusiasts will have wider impact on PEV penetration. The literature includes several other studies conducted in Denmark, the UK, Belgium, Germany, and Netherlands, the overview of which is available in [11]. In Section 13.3, recent approaches to promoting PEV adoption are reviewed.

13.2 Theoretical Frameworks for PEV Adoption

In socio-technical studies, often customer adoption of a new technology is determined via measurable response to an innovation or a new technology [12]. For PEV technology, however, the current adoption rates are low, and instead consumers are evaluated based on their intentions or willingness to purchase a PEV. A variety of theories in economics, psychology, and social studies have been used to explain the factors defining the acquisition of PEVs. The literature is comprised of studies that use normative theories, environmental attitudes, symbols and lifestyle models, and consumer emotions [11]. In this section, we review the details of the most relevant theories that aim to explain consumer choices, and their effects on PEV adoption rate.

13.2.1 Diffusion of Innovations

In order to predict the market share of PEVs, it is critical to account for the role played by innovation and how it affects the sales of new products over time. Diffusion of innovative technologies will take place at different rates over time. The penetration model underlying PEVs can be characterized by innovation diffusion models, which are pioneered by the work of Rogers in 1962 [8]. Rogers argues that innovation must be widely accepted, or reach a tipping point, to sustain itself. The theory of diffusion of innovation has four main components [8]:

- The characteristics of the innovation and how it tackles a product.

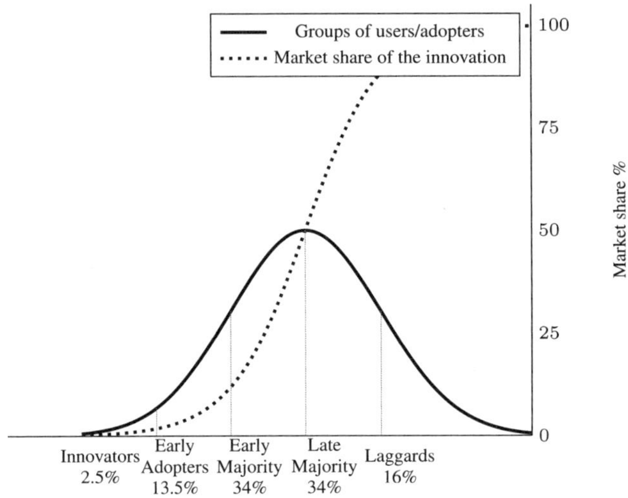

Figure 13.4 The diffusion of innovations model [8].

- Communication channels such as social media, TV, radio, and advertisements that disseminate information about the product.
- Time, which determines at what rate the innovation is propagated.
- Social context, structure, and norms that hinder or facilitate the adoption. For instance, opinion leaders can accelerate the adoption process or have a negative influence on the process.

Furthermore, Rogers characterizes the adopters into five categories as *innovators, early adopter, early majority, late majority*, and *laggards*.

- *Innovators* are the group who are technology enthusiasts, well-informed, and take the risk for unproven products.
- *Early adopters* are also well-informed opinion leaders who have higher social status, a good level of education, and financial liquidity. Their goal is to pioneer the usage of the innovation and be the first to benefit from it.
- *Early majority* are the ones who understand the innovation but do not take risks and want to be confident about the benefits. They mostly rely on the recommendations of other consumers.

- *Late majority* have a high degree of skepticism about the innovation. They also do no have much disposable income to spend for innovations. This group needs to see the obvious benefits to acquire the innovation.
- *Laggards* are the last to adopt an innovation. They are the group of individuals who show no opinion leadership, and oppose the change agents.

As explained above, adopter segments have different psychographic and socioeconomic attributes, which determine their motivation towards purchasing a new product. This means that unidirectional incentives may not be appropriate for each customer group, hence different incentive mechanisms are required for each group. In addition to the aforementioned attributes, customer segments also differ in size. As shown in Figure 13.4, innovators are the smallest group, while early adopters and laggards make up a large group, but not as sizable as the early majority and late majority groups. Due to different sizes of adopter populations, the market share follows an S-shape, as depicted in Figure 13.4. A comprehensive study using the Rogers model has ben conducted in Netherlands in 2012 by using the data collected from almost three thousand respondents [15]. The results for the adopter groups were reported as 9.89%, 42.23%, 38.06%, and 9.82% of the population represent *innovators* (and *early adopters*), *early majority*, *late majority*, and *laggards*, respectively. The Rogers model is also used in many other studies, including [5, 6, 11, 13, 14], to study the future penetration patterns of electric vehicles.

Following Rogers's model, another mathematical model for innovation diffusion is the work of Bass [16]. Bass's model uses differential equations to determine the flow of consumers to a given market out of mutually exclusive and collectively exhaustive subgroups of adopters and non-adopters. Bass characterized innovation diffusion as a spreading process that is established by mass media communication (external influence) and it is moved forward by the word-of-mouth (internal influence). According to the aggregate model approach, innovation diffusion is measured as the flow of adopters $n(t)$ at the time t from a potential market M to the current market. This is often measured as a linear function of an external bias parameter p and an internal influence parameter q [1] (from previous adopters related to $n(t)$, and $N(t)$ is the number of actual adopters at time t). Then the flow of adopters can be calculated according to [14],

$$n(t) = \left(p + q \cdot \frac{N(t)}{M}\right)(M - N(t))\,. \tag{13.1}$$

1 The parameters p and q are seldom referred as coefficients for innovation and imitation, the driving forces of the diffusion pattern.

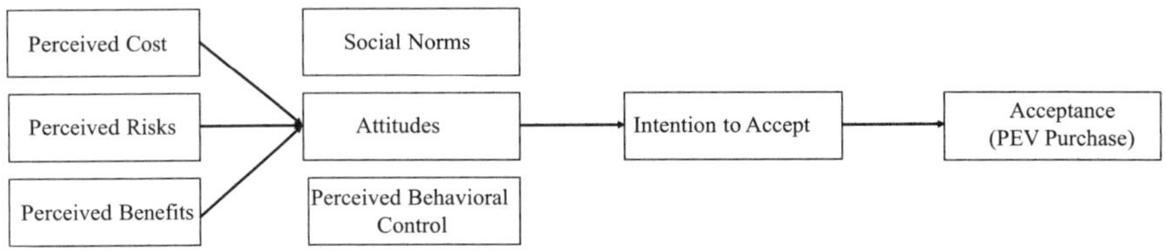

Figure 13.5 Components of the theory of planned behavior (TPB).

Estimating p and q parameters can be challenging for PEV adoption [9]. There is a large body of literature that addresses estimating system parameters [9]. For instance, the work presented in [17] estimates p, q and M for Toyota Prius (hybrid vehicle) as 0.001645368, 1.455182, and 2875826, respectively, in Korea. A more recent study for Denmark is presented in [18], where p and q are estimated to be 0.002 and 0.23, respectively. A global perspective is drawn in [9] in which the market potential of Toyota hybrids in Japan, North America, and Europe is estimated to be 13 million, 4.9 million, and 8.7 million units, respectively. It is noteworthy that for being accurate, the diffusion models require detailed datasets, which is the primary reason underlying the dramatic discrepancies among accuracies.

13.2.2 Theory of Planned Behavior and Rational Choice Theory

The theory of planned behavior (TPB) is one of the best predictor theories deployed for analyzing marketing innovative products [19]. TPB combines the beliefs of the behavior that is predicted by intentions. According to TPB, intentions are created from three distinct processes: innovative behavioral attitudes, subjective norms, and perceived behavioral control. The behavioral attitudes describe how an individual sees and feels about the behavior, and expresses his or her expectations. Behavioral attitudes further show if the change in the attitude is either enjoyable (e.g., noiseless driving) or not necessarily enjoyable but beneficial (e.g., fuel savings and reduced emissions) or harmful. The second predictor, subjective norms, is related to the encouragement given (or not given) by the close social circles of the individual such as family and friends. Subjective norms could be divided into two groups: (1) injunctive norms, which indicate whether an individual's social group stimulates him or her in favor of the behavior (e.g., purchasing a PEV); and (2) descriptive norms, which represent whether the individual's social circles perform the behavior (e.g., purchasing a PEV). The third intention predictor is the perceived behavioral control, which is the perception of how easy or hard it would be to carry out the behavior. For instance, perceived behavioral control could be affected by the presence of chargers, the daily driving ranges, and the duration of charging [20,21]. An overview of TPB is presented in Figure 13.5. TPB has been used in several studies in the literature [10, 20, 21], which study the PEV usage intentions. The studies in [20, 21] collected data from 1202 individuals in Belgium, and argue that consumers are influenced by personal considerations rather than by social factors.

Similar to the theory of planned behavior, rational choice theory (RCT) states that consumers make prudent and rational decisions to maximize their utility for a given set of choices. The discussions presented in Chapters 6, 7, 8, 9, and 11 are related to RCT, as the customers make decisions (e.g., choose a station) to maximize a utility function (e.g., total cost). A handful of studies considered PEV adoption as a rational behavior and measured the customer intentions for using RCT [7, 12, 22, 23]. In these studies, consumers compare PEV attributes such as driving range, speed, initial cost, and operation cost with those of combustion engine vehicles, and form rational decisions. RCT assumes rational consumers and avoids the physiological, emotional, and social factors.

13.3 Case Studies for PEV Promotion

In this section, we provide an overview of policy measures that are used to promote PEV adoption for urban areas. The case for electrification of urban transportation depends on several factors such as automobile density, the role and share of the public transportation system, congestion at roads and parking spaces, and noise and air pollution levels. For instance the share of passenger cars in Shanghai is 20%, while it is almost 80% in Los Angeles. Moreover, in Rotterdam the mode of transportation is dominated by walking, biking, and other urban transport modes. We present an overview of recent innovative solutions, transformative policies, business models, and projects that have been implemented in 23 different countries.

13.3.1 Zones with Controlled Emissions

An ultra low emission zone (ULEZ) is a region in which all vehicles (cars, motorcycles, vans, and buses) should meet emission standards or pay a charge to travel, which might vary depending on the emission of the vehicle. Regulations similar to ULEZ aim to transform a region into PEV-intensive zones [24]. In such zones,

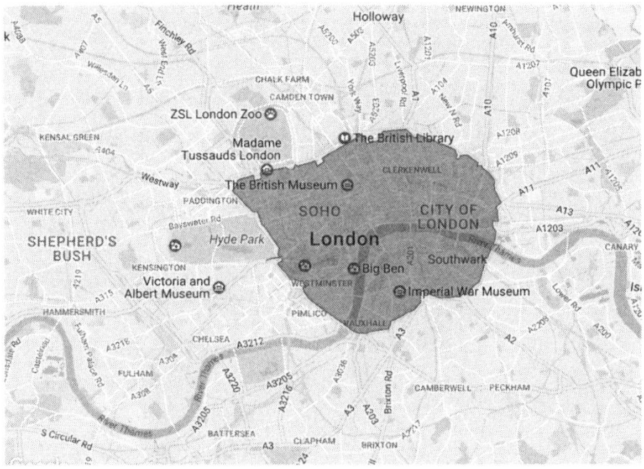

Figure 13.6 Map of the ultra low-emission zone in London [25].

PEVs gain an advantage over ICEVs, as residents of such zones are encouraged to purchase PEVs. Also, the number of low emission zones (LEZ) has been growing across Europe. Specifically, more than 200 cities in 10 countries have already created LEZs or are in the process of launching them, which are also known as Umweltzonen, Milieuzones, Lavutslippssone, Miljozone, and ZTL Ambiente. More information on such zones can be found at the European Commission's Urban Access Regulation website [25]. The maps of low-energy zones are shown in Figures 13.6 and 13.7.

13.3.2 Biotaxis in Bogota, Colombia

Bogota is one of the largest cities in the South America. In Colombia, almost 75% of the population live in urban areas, and taxis are one of the most common transportation options in Bogota. Moreover, the transportation industry is responsible for 44% of the total carbon emissions. In 2013, the city of Bogota launched an initiative to replace high-mileage taxis with electric taxis. This fleet not only reduces the carbon emissions, but also cuts the high maintenance and fuel expenses of the high-mileage vehicles. The project started with 50 vehicles and is expected to reach 2250 electric cars in the near future.

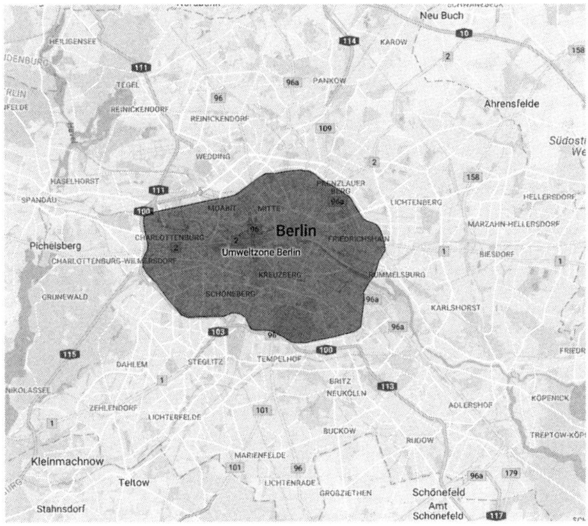

Figure 13.7 Map of the low-emission zone in Berlin, Germany [25].

13.3.3 E-Trike in Manila, Philippines

Asian Development Bank and the Philippine Department of Energy have developed a project to promote three-wheeler electric vehicles. The project is expected to provide cost-savings for low-income communities and to make transportation less sensitive to price fluctuations. The authorities have invested \$500 million to deploy one hundred thousand cars. The project aims to generate \$185 million in savings per year and reduce annual emissions by $400,000$ tons.

13.3.4 Greening Public Transport in Rome, Italy

Even though buses are an efficient means of public transportation, they are also the major sources of urban air pollution. The city of Rome has one of the largest bus fleets, which serves more than nine hundred million passengers annually. Local administrators created limited access zones exclusively for electric buses, which resulted in a reduction in emissions and noise. The studies estimate that energy consumption of electric buses is 65% less than hybrid buses and 75% less than diesel buses.

13.3.5 Car-sharing in Paris, France

Autolib company has introduced an electric vehicle-sharing service in Paris, France. The sharing network includes about four thousand vehicles and six thousand charging points. Car sharing has proven to be an effective solution to increased costs of car ownership and limited availability of dedicated parking spaces in large metropolitan areas like Paris. Shared PEVs also means that individuals do not face high purchase cost and battery degradation of batteries. The car-sharing system has received considerable attention and Autolib is expanding its network in other cities like Lyon, Bordeaux, London, and Indianapolis.

13.3.6 Workplace Charging Challenge, United States

As a part of the Department of Energy's EV Everywhere initiative, the Workplace Charging Challenge program aims to promote the deployment of PEV chargers at workplaces. The ability to charge during work hours can double the daily driving range of electric miles. More than 300 companies have joined the program. As of 2014, 80% of the employees are provided with the service for free, and on average, there are 11 chargers and 26 PEVs in each workplace. Moreover, workplaces offering charging services promote PEV adoption. In workplaces that offer charging services, the employees are twenty times more likely to drive a PEV than the average employee [25].

13.3.7 Zero-Emission Vehicle Program, United States

The zero-emission vehicle (ZEV) program is managed by the California Air Resources Board, and it is accepted in nine states (Connecticut, Maine, Maryland, Massachusetts, New Jersey, New York, Oregon, Rhode Island, and Vermont) to boost PEV (both BEV and PHEV) sales. Car manufacturers are obliged to sell a certain number of PEVs in these states. Each automaker is assigned ZEV credits that represent PEV sales. Moreover, due to various driving ranges, each PEV model is represented with a different credit rating. For example, a Tesla with a 200-mile electric range represents 3.3 ZEV credits, while a Nissan Leaf with 84-mile electric range represents 1.8 PEV credits. Similarly, a PHEV with 10-mile electric range deserves 0.1 credit. In 2018, the ZEV credit requirement is set as 4.5%, while it is set to 22% in 2025. For example, if a car manufacturer sells 100,000 vehicles, the company needs to collect 4,500 ZEV credits, where at least 2,000 should come from PEVs.

In addition to the programs described above, municipalities and utilities further offer benefits for PEV adoption, such as unrestricted HOV lane access, exemption from inspections, free parking, free charging, and expatiated permit process for home charger installation. The work presented in [27] takes all of the offered PEV incentives (also the ones presented in Chapter 3) and ranks the major United States cities according to their readiness for PEV usage. The top ten cities are ranked as Portland, OR, Washington, DC, Baltimore, MD, New York City, NY, Denver, CO, Los Angeles, CA, Atlanta, GA, Chicago, IL, Austin, TX, and San Francisco, CA. It is noteworthy that most of the top cities are located in states that use the ZEV program.

13.4 Chapter Summary

In this chapter, we presented some socioeconomic and policy-related perspectives associated with PEV adoption. It is evident that current financial incentives and subsidies may have only limited effects on PEV adoption if consumers have little confidence in PEV awareness. To shed light on this, we provided a detailed overview of some of the recent field studies and concluded that PEV cost is still the major obstacle for most drivers. Furthermore, we presented theoretical frameworks from behavioral economics that study how PEVs are disseminated. Experience from past technology adoption suggests that PEVs are naturally attractive to a few early adopter demographics, such as technology enthusiasts and visionaries, while it takes several years for a significant portion of the population to adopt PEVs. As a result, policy makers should focus on early adopters rather than mainstream customers to boost the PEV sales. The last section of the chapter highlighted several programs that were designed to promote PEVs in various regions across the globe.

References

[1] Kempton, W., "Electric Vehicles: Driving Range," *Nature Energy*, Vol. 1, 2016, pp. 16131.

[2] Needell, Z. A., et. al., "Potential for Widespread Electrification of Personal Vehicle Travel in the United States," *Nature Energy*, Vol.1, 2016, pp. 16112.

[3] Kezunovic, M., Waller, S. T., and Damnjanovic, I., "Framework for Studying Emerging Policy Issues Associated with PHEVs in Managing Coupled Power and Transportation Systems," *Proc. IEEE Green Technologies Conference*, 2010, pp. 1-8.

[4] Singer, M.,"Consumer Views on Plug-in Electric Cehicles, '*National Benchmark Report (No. NREL/TP–5400-65279). NREL (National Renewable Energy Laboratory (NREL)*, Golden, CO (United States), 2016.

[5] Green, E. H., Skerlos, S. J., and Winebrake, J. J., "Increasing Electric Vehicle Policy Efficiency and Effectiveness by Reducing Mainstream Market Bias," *Energy Policy*, Vol. 65, 2014, pp.562-566.

[6] Sierzchula, W., et.al, "The Influence of Financial Incentives and Other Socio-economic Factors on Electric Vehicle Adoption," *Energy Policy*, Vol. 68, 2014, pp. 183-194.

[7] Carley, S., et. al., "Intent to Purchase a Plug-in Electric Cehicle: A Survey of Early Impressions in Large United States Cites," *Transportation Research Part D: Transport and Environment*, Vol. 18, 2013, pp. 39-45.

[8] Rogers, E. M. (2010). *Diffusion of Innovations*, 2010, Simon and Schuster.

[9] Massiani, J., and Gohs, A., "The Choice of Bass Model Coefficients to Dorecast Diffusion for Innovative Products: An Empirical Investigation for New Automotive Technologies," *Research in Transportation Economics*, Vol. 50, 2015, pp. 17-28.

[10] Egbue, O., and Long, S., "Barriers to Widespread Adoption of Electric Vehicles: An Analysis of Consumer Attitudes and Perceptions ," *Energy Policy*, Vol. 48, 2014, pp. 717-729.

[11] Rezvani, Z., Jansson, J., and Bodin, J., "Advances in Consumer Electric Vehicle Adoption Research: A Review and Research Agenda ," *Transportation research part D: transport and environment*, Vol. 34, 2015, pp. 122-136.

[12] Jensen, A. F., Cherchi, E., and Mabit, S. L., "On the Stability of Preferences and Attitudes Before and After Experiencing an Electric Vehicle ," *Transportation Research Part D: Transport and Environment*, Vol. 25, 2013, pp. 24-32.

[13] Peters, A., and Dutschke, E. (2014). "How Do Consumers Perceive Electric Vehicles? A Comparison of German Consumer Groups," *Journal of Environmental Policy & Planning*, Vol. 16, No. 3, 2014, pp. 359-377.

[14] Kiesling, E., et. al., "Agent-based Simulation of Innovation Diffusion: A Review," *Central European Journal of Operations Research*, Vol. 20, No. 2, 2012, pp.183-230.

[15] Bockarjova, M., Rietveld, P., and Knockaert, J., "Adoption of Electric Vehicle in the Netherlands-A Stated Choice Experiment", *Tinbergen Institute Discussion Paper*, 2013.

[16] Bass, F. M., "A New Product Growth for Model Consumer Durables," *Management science*, Vol. 15, No. 5, 1969, pp. 215-227.

[17] Park, S. Y., Kim, J. W., and Lee, D. H., "Development of a Market Penetration Forecasting Model for Hydrogen Fuel Cell Vehicles Considering Infrastructure and Cost Reduction Effects," *Energy Policy*, Vol. 39, No. 6, 2011, pp. 3307-3315.

[18] Jensen, A. F., et. al.,"Predicting the Potential Market for Electric Cehicles," *Proc. In Transportation Research Board 93rd Annual Meeting*, No. 14-3855, 2014.

[19] Ajzen, I., "The Theory of Planned Behavior," *Organizational behavior and human decision processes*, Vol. 50, No. 2, 1991, pp. 179-211.

[20] Moons, I., and De Pelsmacker, "Emotions as Determinants of Electric Car Usage Intention," *Journal of Marketing Management*, Vol. 28, No. 3, 2012, pp. 195-237.

[21] Moons, I., and De Pelsmacker, P., "An Extended Decomposed Theory of Planned Behavior to Predict the Usage Intention of the Electric Car: A Multi-group Comparison," *Sustainability*, Vol.7, No. 5, 2015, pp. 6212-6245.

[22] Krupa, et. al.,"Analysis of a Consumer Survey on Plug-in Hybrid Electric Vehicles," *Transportation Research Part A: Policy and Practice*, Vol. 64, 2014, pp. 14-31.

[23] Lieven, T., Muhlmeier, S., Henkel, S., and Waller, J. F., "Who will buy electric cars? An empirical study in Germany," *Transportation Research Part D: Transport and Environment*, Vol. 16, No. 3, 2011, pp. 236-243.

[24] EV City Casebook: 50 Big Ideas Shaping the Future of Electric Mobility, 2014, International Energy Agency .

[25] Urban Access Regulation in Europe http://urbanaccessregulations.eu/.

[26] U.S. Department of Energy's EV Everywhere Workplace Charging Challenge Progress Update 2014: Employers Take Charge, http://www.afdc.energy.gov/uploads/publication wpc_2014_progress_report.pdf.

[27] Clark-Sutton, K. et. al., "Plug-in Electric Vehicle Readiness: Rating Cities in the United States," *The Electricity Journal*, Vol. 29, No.1, 2016, pp. 30-40.

List of Acronyms

BEV Battery electric vehicle

HEV Hybrid electric vehicle

ICEV Internal combustion engine vehicle

IEEE Institute of Electrical and Electronics Engineers

ISO Independent system operator

PEV Plug-in electric vehicles

PHEV Plug-in hybrid electric vehicle

QoS Quality of service

RTO Regional transmission operator

SAE Society of Automative Engineers

SoC State of charge

SoH State of health

V2B Vehicle-to-building

V2G Vehicle-to-grid

V2H Vehicle-to-home

V2V Vehicle-to-vehicle

ZEV Zero electric vehicle

About the Authors

İslam Şafak Bayram is an assistant professor at Division of Sustainability at College of Science and Engineering and a scientist at Qatar Environment and Energy Research Institute, both at Hamad Bin Khalifa University. He received a B.S. degree in electrical and electronics engineering from Dokuz Eylul University, Izmir, Turkey in June 2007, an M.S. degree in telecommunications from the University of Pittsburgh in August 2010, and a Ph.D. degree in computer engineering from North Carolina State University in December 2013. He received the Best Paper Award at the Third IEEE International Conference on Smart Grid Communications and at the First IEEE Workshop on Renewable Energy and Smart Grid in March 2015. Between January-December 2014, he worked as a postdoctoral research scientist at Texas A&M University at Qatar. His research interests include stochastic processes, queueing theory, optimization, economic modeling and their applications to plug-in electric vehicles, demand-side management, smart power grids, and communication networks.

Ali Tajer is an assistant professor of electrical, computer, and systems engineering at Rensselaer Polytechnic Institute. During 2007-2010 he was at Columbia University where he received an M.A degree in Statistics and a Ph.D. degree in electrical engineering, and during 2010-2012 he was at Princeton University as a postdoctoral research associate. His research interests include mathematical statistics and network information theory, with applications in wireless communications and power grids. Dr. Tajer serves as editor for the *IEEE Transactions on Communications*, editor for the *IEEE Transactions on Smart Grid*, guest editor for the *IEEE Signal Processing Magazine*, and previously as guest editor-in-chief for the *IEEE Transactions on Smart Grid* Special Issue on "Theory of Complex Systems with Applications to Smart Grid Operations." He is a senior member of the IEEE and received an NSF CAREER award in 2016.

Index

Achievable cycle count (ACC), 68–70
Admission control, 206–208
All-electric range (AER)
    AER, 9, 12, 14, 15, 57, 98
Ancillary services, 227–231, 234, 235, 243
Andreas Flocken, 5
Argonne National Laboratory, 16, 243
Arrhenius relationship, 64
Arrival rate, 56, 139–142, 145–150, 152, 156–158, 207–211, 213, 220
Automatic generation control (AGC), 227, 229
Availability factor, 228, 229

Bass's model, 253, 260
Battery
    C-rate, 26
    cell, 24, 25, 30, 31, 36, 61, 71, 72, 168
    Charge depleting, 28
    Charge sustaining, 28
    Cut-off voltage, 28
    Cycle life, 26, 29, 30, 43, 62, 64, 65
    Degradation, 26, 30, 36, 41, 61–65, 162, 163, 169, 176, 177, 180, 239, 258
    Depth of discharge (DoD), 26, 64, 67–70
    Energy density, 1, 7, 27, 29, 30
    Energy rating, 100, 138
    Float voltage, 27
    Internal Resistance, 28, 63–66, 69
    lead-acid, 6
    Li-ion, 24, 25, 31, 61, 62, 68, 69, 80, 161, 163, 164, 168, 169
    Lithium-ion, 24, 68
    module, 25, 112, 113
    power density, 27
    Specific energy, 27
    Specific power, 27, 30
    State of charge (SoC), 26–29, 31, 63, 66–69, 161–166, 169–173, 176, 177, 179, 180, 237
    State of health (SoH), 26, 31
Battery electric vehicles (BEVs), 8–16, 20, 25, 28, 32–34, 250
Battery management systems, 31

Blocking probability, 44, 99, 120, 121, 125–128, 133, 136, 137, 142, 145, 148, 149, 151, 155, 170, 174, 177, 179, 180, 209, 211, 213, 216, 221

California Independent System Operator (CAISO), 229, 230, 235, 236, 243
Camille Jenatzy, 5
Capacity planning, 56, 97, 139
Charging rates
  CHAdeMO, 34, 39, 40
  DC fast, 35, 37–39, 44, 55, 56, 71, 97, 119, 155–158, 205, 206, 212, 223, 235
  Level 1, 37–39, 41, 98, 109, 114, 119, 155, 185, 196
  Level 2, 37–39, 41, 97, 109, 114, 155, 185
  Level 3, 38
  Tesla SuperCharger, 34, 40
Constant current constant voltage (CCCV), 161, 163, 168, 169, 180
Consumer average interruption duration index (CAIDI), 241
Consumer innovativeness, 251
Critical peak pricing, 74, 75
Crude oil, 1, 5, 257
  oil market, 2, 3, 7

Demand charges, 73, 74, 237–240, 242
Diffusion of innovations, 45, 246, 251, 252, 260
Doyle-Fuller-Newman, 71

Economic control, 232
Effective demand, 110, 111, 114–116
Electric Reliability Council of Texas (ERCOT), 230, 232, 233
Electric vehicle service equipment (EVSE), 34, 35, 40, 42
Energy storage systems (ESS), 98, 99, 101, 103, 121, 123, 126–128, 133, 135, 137
Environmental Protection Agency (EPA), 2, 10–12, 15

Facility location, 184–186, 188, 189, 192, 193, 198, 201–203
Flow refueling location model (FRLM), 197, 198, 202
Flow-capturing location problems (FCLPs), 196
Frequency bias, 231
Frequency regulation, 229, 230, 232, 235, 242
Fuel-cell electric vehicle (FCEV), 8, 10, 20

Gaston Plante, 5
Greenhouse gas
  GHG emissions, 2–4, 13

Harmonic distortion, 46, 48, 49, 59
Hessian matrix, 136
Hotelling law, 217
Hybrid electric vehicles (HEV), 9, 10, 15
Hybridization factor, 34

Inertial response, 231
Internal combustion engine vehicle (ICEV), 2, 3, 5–7, 9–11, 15, 32, 34, 74, 248, 255

International Council on Large Electric Systems (CIGRE), 51
International Electrotechnical Commission (IEC), 38, 39, 51, 52
IEC 1000-3-2, 37
IEC 61851, 38
IEC 61851-23, 39
IEC TC69, 38

Jacobian, 136

Kaufman-Roberts, 153, 155

Little's law, 175
Logit transformation, 135
Loss of load probability, 147–155, 157, 206, 208

M/D/C, 174
M/G/C, 174
M/M/C, 174
M/M/C/C, 141, 174
Markov chain, 106, 109, 110, 123–126, 132–134, 151
Maximum covering, 186, 190, 202
Midwest ISO, 230
Miles per gallon equivalent (MPGe), 10–12
Monroney label, 11

National Electrical Manufacturers Association (NEMA), 51
National Household Transportation Survey, 229
Network externalities, 93
New England ISO, 230
New York ISO, 230

Ontario ISO, 230

P-center, 186, 190, 191, 193, 202
P-median, 187, 190, 202
Photovoltaic (PV), 240, 242
PJM (Pennsylvania, Jersey, Maryland) RTO, 226, 230, 232–235
Plug-in electric vehicle (PEV), 2, 8, 10–13, 16–21, 23–25, 27–31, 34–42, 45–55, 57–59, 61, 62, 68, 70, 71, 75, 78, 81, 97–104, 106, 108–112, 114, 117, 119–122, 127, 133–135, 139, 141–143, 145–150, 152, 154, 157, 158, 162–164, 169, 170, 172, 175, 176, 179, 181, 183–186, 188, 193, 195–197, 199, 202, 203, 205–210, 213–217, 220–222, 225–242, 245–251, 254–256, 258–260
Plug-in hybrid electric vehicle (PHEV), 8, 9, 15, 21, 180, 249, 261
Poisson process, 100, 122, 141, 148, 152, 170
Power grid, 2, 3, 9, 10, 23, 34, 36, 40, 46, 55, 59–61, 97–99, 141, 147, 205, 212, 226–228, 240, 241

Quality of service, 91, 99, 128, 146, 162, 207–211, 213–216

Rational choice theory, 246, 255
Real-time pricing, 75
Regional transmission operators (RTOs), 230
Regulation control, 231

Reserves, 230
  Non-spinning, 233
  Spinning, 233, 242
Response surface methodology (RSM), 135
Robert Anderson, 5
Rogers model, 251–253

Smog, 15
Smog rating, 14
Social welfare, 92, 141, 146, 148, 149, 156–158
Socially optimal, 88–90, 93
Society of Automative Engineers (SAE), 34, 37
  SAE J1772, 40
  SAE J2836/1, 40
  SAE J2836/2, 40
  SAE J2836/3, 40
  SAE J2836/4, 40
  SAE J2836/5, 40
  SAE J2836/6, 40
  SAE J2847/1, 40
  SAE J2847/2, 40
  SAE J2847/3, 40
  SAE J2847/4, 40
  SAE J2847/5, 40
  SAE J2847/6, 40
Stackelberg game, 206, 214, 215, 221
  Follower, 206, 214, 215, 221
  Leader, 206, 214, 217, 221
System average interruption duration index (SAIDI), 241
System average interruption frequency index (SAIFI), 240, 241
System losses, 50, 59, 148, 162

Theory of planned behavior, 261
Theory of planned behavior (TPB), 246, 254, 255
Thermal loading, 46, 47, 59
Time of Use, 74, 78, 127, 163, 240

United States Advanced Battery Consortium LLC (USABC), 29, 30
Utility function, 86, 88, 91, 92, 140, 148–151, 156, 206, 214, 216, 217, 255

Vehicle-to-building (V2B), 225, 226, 237–241
Vehicle-to-grid (V2G), 225–229, 235, 237–239, 242
Vehicle-to-home (V2H), 225, 226, 239–241
Vehicle-to-vehicle (V2V), 225, 226, 239, 241, 242
Voltage
  Deviation, 53
  Drops, 233
  Profile, 52, 71, 81, 162, 168
  Sags, 163, 233
  Voltage source, 163

Weber problem, 189
Well-to-wheel, 11, 15, 17
William Morrison, 5
Wireless charging, 40, 41, 44

Zero-emission vehicle, 258, 259

Recent Artech House Titles in Power Engineering

Andres Carvallo, Series Editor

Advanced Technology for Smart Buildings, James Sinopoli

The Advanced Smart Grid: Edge Power Driving Sustainability, Second Edition, Andres Carvallo and John Cooper

Battery Management Systems, Volume I: Battery Modelings, Gregory L. Plett

Battery Management Systems for Large Lithium Ion Battery Packs, Davide Andrea

Battery Power Management for Portable Devices, Yevgen Barsukov and Jinrong Qian

Design and Analysis of Large Lithium-Ion Battery Systems, Shriram Santhanagopalan, Kandler Smith, Jeremy Neubauer, Gi-Heon Kim, Matthew Keyser, and Ahmad Pesaran

Designing Control Loops for Linear and Switching Power Supplies: A Tutorial Guide, Christophe Basso

Electric Power System Fundamentals, Salvador Acha Daza

Electric Systems Operations: Evolving to the Modern Grid, Mani Vadari

Energy Harvesting for Autonomous Systems, Stephen Beeby and Neil White

GIS for Enhanced Electric Utility Performance, Bill Meehan

Introduction to Power Electronics, Paul H. Chappell

Introduction to Power Utility Communications, Dr. Harvey Lehpamer

IoT Technical Challenges and Solutions, Arpan Pal and Balamuralidhar Purushothaman

Battery Management Systems, Volume I, Battery Modelings, Gregory L. Plett

Plug-in Electric Vehicle Grid Integration, Islam Safak Bayram and Ali Tajer

Power Line Communications in Practice, Xavier Carcelle

Power System State Estimation, Mukhtar Ahmad

A Systems Approach to Lithium-Ion Battery Management, Phil Weicker

Signal Processing for RF Circuit Impairment Mitigation in Wireless Communications, Xinping Huang, Zhiwen Zhu, and Henry Leung

Synergies for Sustainable Energy, Elvin Yüzügüllü

Telecommunication Networks for the Smart Grid, Alberto Sendin, Miguel A. Sanchez-Fornie, Iñigo Berganza, Javier Simon, and Iker Urrutia

For further information on these and other Artech House titles, including previously considered out-of-print books now available through our In-Print-Forever® (IPF®) program, contact:

Artech House
685 Canton Street
Norwood, MA 02062
Phone: 781-769-9750
Fax: 781-769-6334
e-mail: artech@artechhouse.com

Artech House
16 Sussex Street
London SW1V 4RW UK
Phone: +44 (0)20 7596-8750
Fax: +44 (0)20 7630-0166
e-mail: artech-uk@artechhouse.com

Find us on the World Wide Web at: www.artechhouse.com